Making Technology Masculine

Making Technology Masculine

Men, Women and Modern Machines
in America, 1870-1945

Ruth Oldenziel

AMSTERDAM UNIVERSITY PRESS

Cover illustration: Cover of the first issue of *The Californian Journal of Technology*, 1,1 (February 1903). Courtesy of University of California Library, Berkeley, CA.

Cover design: Neno Ontwerpers, Amsterdam
Lay-out: JAPES, Amsterdam
ISBN 90 5256 381 4
© Amsterdam University Press, Amsterdam, 1999

Table of Contents

Acknowledgements

Scholars in the fields of American Studies, Women's Studies, and History of Technology helped shape and reinforce my commitment to interdisciplinary research. Nancy F. Cott, Jean-Christophe Agnew, and David Montgomery left the intellectual space and asked the analytically essential questions; all three carefully read and engaged with earlier drafts. Since then, I found many communities of thinkers.

The research of the book was funded by fellowships from Yale University, the Smithsonian Institution, the Eleutherian Mills-Hagley Foundation, the Hoover Presidential Library, and the Lemelson Center. Most of my research skills were learned from scores of reference librarians as they shared their knowledge during a time when computer technologies were rapidly changing the content of their jobs and the nature of scholarly research. Jim Ronan at the National Museum of American History and the reference staff at Yale Sterling Memorial Library, in particular, helped find creative solutions to challenging questions. From them I acquired my appreciation for the variables in the organization and classification of knowledge. Another three hundred reference librarians, officers, and volunteers at university archives, engineering colleges, and alumni organizations in often underfunded and understaffed offices, answered my survey on women in engineering, checking and rechecking their records; many of whom I got to know through the mail; among them Margaret Raucher at the Walter P. Reuther Library in Detroit deserves special mention.

Many scholars also answered my queries: I am grateful to Ava Baron, David Hounshell, Peter Liebhold, Steve Lubar, Art Molella, David Noble, Margaret Rossiter, John Staudenmaier, Carlene Stephens, Jeffrey Stine, Martha Moore Trescott, Matt Turner, Deborah Warner, and Bill Worthington. Since then I have found inspiration in the many conversations with colleagues who became friends: Nina Lerman, Arwen Mohun, Carolyn Goldstein, Carolien Bouw, Irene Cieraad, and José van Dijck. My thanks also go to the colleagues who took their precious time and read the entire manuscript: Pat Cooper, Frances Gouda, Rosa Knorringa, and Phil Scranton. Portions were read by Eric Schatzberg, Ronald Kline (chapter 1), Liesbeth Bervoets (chapter 2), Bruce Sinclair, Ton Brouwer, José van Dijck (chapter 3), and Karin Zachmann (chapter 6). I am most grateful to Joe Corn and Arwen Mohun who read it more than once asking critical questions and providing substantive comments.

José Quiroga, Frances Gouda, and Gary Price offered a pied-a-terre in Washington providing an intellectual hiding place throughout the years, for which I am deeply grateful. In the finishing stages, the Gietelink and Verster families stood always ready to welcome Alexander in their midst. I could not have travelled this journey without the friendship of Rosa Knorringa and Frederike de Vlaming, however. They were the most persistent supporters whose trust engendered trust in the project and whose love and friendship have enriched me in the past decades. I dedicate the book to them.

Introduction

Men's love affair with technology is something we take for granted. Only when that affair runs amok is it likely to elicit any serious commentary at all. The editors of *The New York Times* considered the subject of male technophilia sufficiently newsworthy to devote a front-page article to it in 1986 when Robert Morris, Jr., generated "the biggest computer gridlock" ever known. His program jammed over 6000 computers in the United States, including some in the military computer network. Women and girls "use computers; men and boys love them. And that difference appears to be a critical reason that computers in America remain a predominantly male province," the article declared. It went on to report that women "are almost without exception bystanders in the passionate romance that men conduct with these machines, whether in computer science laboratories, video game parlors, garages or dens."

Not only journalists recognized the gendered aspects of the computer incident. When commenting on his son's obsession with computers, the student's own father, one of the government's most respected computer security experts, wryly remarked: "I had a feeling this kind of thing would come to an end the day he found out about girls." He added, "Girls are more of a challenge." Both the *Times* journalist and Morris Sr. suggested that a widespread male cultural pattern of passion with technology had endangered America's military and national security interests. If taken to the extreme, that passion might have dire consequences, indeed. Most commentators paid attention to the effects rather than to the underlying source of that love affair: after all, men's attraction to technology was, and still is, considered a matter of fact that needs no further explanation.

Whenever women enter computer rooms and construction sites as designers, hackers, and engineers, however, they need to be accounted for and explained. For decades scores of newspapers have reported, commented, and elaborated on the many "first" women who trespassed the male technical threshold as engineers, presenting them often as news. During the nineteenth century Emily Roebling (1841-1902), her husband's business partner during his twenty-year-long illness, elicited commentary when she helped supervise the construction of New York City's Brooklyn Bridge. For years, Kate Gleason (1865-1933) was touted to be the first women engineer even though many lesser known women were also working in the field at the time. No doubt responding to the governmental campaigns to recruit women as technical personnel, *The Christian Science Monitor* claimed in 1943 that Isabel Ebel was "the only woman aeronautical engineer of the United Airlines staff." Even after the war emergency, *The New York Times* considered the graduation of Audrey Muller newsworthy enough to call her "the first woman in the history of the University of Michigan to receive a bachelor in naval architecture." In the home town of General Electric Co., an employer of scores of women in technical positions for many decades, *The Schenectady Gazette* still claimed in 1960 that Leonore Traver was just one of the very few women in the country pursuing a career in civil engineering, ignoring the thousands who had preceded her. And as late as 1970, a Philadelphia city paper celebrated Candace Martin, "the first woman member of the Local 4 E International Union of Operating Engineers, [who is] a field engineer on the construction site."[2] From New York to California engineering journals, corporate newsletters, and local newspapers singled out women who trespassed on the male domain of engineering, often adding local touches and highlighting them with photographs to suit the particular occasion. The publicity on women engineers – one might even call it overexposure – shows how we continue to view their entry into the technical domain as an exotic but more likely an exceptional, strange, and alien event. It also illustrates how we forget, erase, and (re)invent the history of women. More importantly, these reports show how we consider technology men's natural domain – a penchant that does need explanation, however.

Morris' male romance with technology has a history, albeit it a short one. There is nothing inherently or naturally masculine about technology. The representation of men's native and women's exotic relationship with technology elaborates on a historical, if relatively recent and twentieth-century Western tendency to view technology as an exclusively masculine affair. The public association between technology and manliness grew when male middle-class attention increasingly focussed its gaze on the muscular bodies of working-class men and valorized middle-class athletes, but disempowered the bodies of Native Americans, African Americans, and women. Similarly the erasures of workers, Native Americans, African Americans, and women from the technical domain was not accidental. This occurred when scores of working-class women entered the labor market and confronted new machines in their jobs as cigar makers, secretaries, switchboard operators, and dress makers; when middle-class women organized both inside and outside the women's movement to stake out new terrains and implement new agendas; and when educated middle-class women sought access to literature, chemistry, medicine, and law, shaping new professional identities. The links between technological change and gender relations developed neither in isolation nor independently. Instead, they shaped each other. In the cultural grammar of the twentieth century, the simultaneous erasure and overexposure of flesh-and-blood women engineers like Ebel, Muller, and Traver evolved together with the shaping of a new technical world inscribed as male.

Making Technology Masculine explores the historical origins of Morris Jr.'s love affair with computer networks in which women had no place; it traces the development of a masculine mystique with and of a female fear of technological change in the last hundred years; it examines how American engineers began to lay claim to a new knowledge domain they called technology while making universal claims for it; it describes how American engineers and their allies employed discourse, language, and narrative strategies and practiced a style of engineering that came to support this gendered division of cultural labor. In these settings, women entered the scene actively plotting their own narratives around fictional engineers to counter their male colleagues on the subject as professional women writers; they turned up insisting on taking their place in the technical arena as engineers. They also appeared silenced as wives and daughters in the autobiographical accounts male engineers constructed. The book is thus a venture that goes beyond a narrative of women's participation in or exclusion from past technological undertakings to chart how notions of gender and technology construct each other. It treats the absence of flesh-and-blood women in technological matters in its relation to their persistent and haunting metaphorical presence.

Men, too, quite emphatically enter these pages at work in their relationships with both women and other men different in class and ethnicity. They enter the stage in search of their own version of male identity as professionalizing engineers looking for cultural resources to upgrade their occupation; as struggling rank-and-file members living in fear of being declassed and demasculinized; and as writers, visual artists, and social scientists in search of their own professional identities by electing engineers as their new models of white manliness and charting a revitalized male identity for the middle class. This book therefore does not deal with women simply as the exclusive bearers of gender. It focuses its gaze on men as gendered male as they shape their stories, professional strategies, and identities.[3]

Focussing one's gaze on men helps to understand why technology developed into a powerful symbol of male, modern, and western prowess; how machines like cars, bridges, trains, and planes have become the measures of men, from which women have been excluded as a matter of course; why corsets have been banished to the basements of the modern classification systems of technology; why women – when they do appear on the scene as engineers and inventors – function like *deae ex machina*. Like the Greeks, who used dramatic devices to lower their gods onto the stage by a machine, our contemporary mythologies often produce women as goddesses whose lives are essentially off-stage, who appear to come from nowhere, and whose plots are engineered elsewhere. In this construction, women who enter the male-defined technical stage always look like amateurs.

Engineers and their advocates were important actors in the construction of such a plot. They came to be seen as the exclusive bearers of a domain we have come to call technology – both as an intellectual construct and material practice. They emerged as the shock troops of modernity from the tracks deep into the western territories, the bottom of South African mines to the basements of New York drafting departments and the laboratories of General Electric bringing good things to life. Between 1870 and 1950, the number of engineers grew 17 times faster than the U.S. labor force as a whole. In the nineteenth century they specialized in civil, mining, and mechanical engineering; at the end of the century the chemical, electrical, and aeronautical industries demanded new kinds of engineering skill and knowledge. Engineers could be found in any function, industry, or geographical location, working for local, municipal, state, or federal governments, for large corporations, or as independent operators. Engineers helped organize capital, hired labor, calculated estimates, signed contracts or carried out research. Since the end of the nineteenth century, American engineering has expanded the most rapidly of all occupations, reflecting the growth and consolidation of modern industry and the spectacular expansion of middle management in the emerging corporations and federal bureaucracies. It was also transformed from an elite profession into a mass occupation. Moreover, it became a segmented and divided profession: by 1935 there were 2,518 different job titles under its rubric. Few could call themselves chief or consulting engineers; most worked in poorly-lighted, crowded, and dirty drafting departments tracing, detailing, lettering, checking, and copying maps, grade profiles, steel structures, plant lay-outs, and underground mine surveys.[4]

In America engineers thus belonged to a deeply divided, segmented occupation. It lacked such classic gate-keeping mechanisms of professionalization as credentialling and licensing. It also lacked a clear-cut identity – an irony, to say the least, given the role engineers were to play in the modern meanings of technology. Despite its relative open, eclectic, and segmented character American engineering remained the most male dominated of all. The number of women who received engineering degrees increased overall, but it did not increase proportionally: they consistently accounted for about three percent of the profession during the first half of the century or probably a few thousand women employed in engineering jobs in the United States by the 1920s, and about six thousand by the end of the second World War.[5]

Why so few women figured in engineering is asking the wrong question, however. Aggregate figures are a matter of definition. An exclusive attention to figures tends to blame women for their inadequate socialization and to ignore the profes-

sional politics behind the creation of such statistics. Government statistics and engineering school records fail to do justice to the daughters and wives who acquired
technical knowledge informally through family firms without ever attending a
specialized school, to those who obtained engineering positions through corporate
on-the-job training after completing a science education, to the daughters of
lower-class and immigrant families who attended evening classes in the hope of
improving their chances for career advancement, or to the hundreds of thousands
of women who trained in engineering during the wars.[6] More importantly, any estimates on the number of women engineers depend highly – if not primarily – on
one's definition of engineering. Should it include the underbelly of the profession:
lab assistants, draftsmen, chemists, detailers, checkers, tracers, and testing technicians? These are hardly innocent choices for these definitions also have a history.
Census figures have been subject to the changing definitions of engineering, reflecting not only statisticians' desire for more precision, but more importantly, the
profession's aspirations for higher status by barring more and more groups previ-
ously included in the definition. These attempts at professionalization often
helped in reinscribing and maintaining engineering as a male middle-class domain
in the course of the twentieth century. Such attempts to upgrade the profession involved cultural work.

Engineers built bridges. They also constructed cultural infrastructures and engaged in narrative productions.[7] Strategies of professionalization, the compilation
of encyclopedias, the writing of autobiographies, the singing of songs, and the telling of jokes were all part and parcel of the cultural work of maintaining engineering as male occupation. As engineering transformed to a mass profession, students
at the leading engineering school MIT proudly, if jokingly, reestablished the male
premise of their profession when confronted with a few women on campus. In a
boisterous and rowdy mood, the male engineering students joked with their female colleagues in a 'co-ed' song entitled *A Son (?) of the M.I.T.* composed in 1907.
They sang, "I would not be a Yale man, Reformers to annoy. Nor yet a Harward
student [sic]: defeat I don't enjoy...Such models I'd not choose....[but] I'm a son
of the M.I.T." Alternatively – and this is the humorous point of the song – the few
women who at the time attended the engineering school would interject, "I'm *not*
a son of the M.I.T." In the narrow space of one eighth note, "I'm *not*," they sang,
"certainly not, and I'm glad of it!" or, "the idea is preposterous!"[8] That eighth note
in the musical phrase represents the narrow space allotted to women entering the
engineering profession since as a point of entry it did not allow for passage into the
bastion. Humor, a grateful tool, helped to relieve the tensions that were part and
parcel of a society that opened its doors to new social groups. For these male and
female voices were engaged in more than mere incidental banter. The songs
showed the kind of negotiation in which men and women were engaged at the

turn of the century when older engineering elites confronted scores of sons of immigrants, some African Americans, and a number of women starting to demand
their rightful place. The entry of immigrant sons through the system of American
engineering education required a lot of work to keep class within its bounds. They
helped to revitalize the workings of middle-class manliness when boundaries of
class were under negotiation and to recast the profession as a middle class, white,
and male jurisdiction by the end of the nineteenth century. In these settings, racial
and gender lines were therefore more carefully drawn than those of class. Like all
jokes, they released a host of tensions, and with them, new meanings in a society
that sought to negotiate the challenges posed by many of its citizens who were
striving to be part of the polity.

Historians employ the nineteenth-century term "useful arts" and the modern
word "technology" interchangeably, as if they were synonymous.[9] As the history of
the term technology suggests, however, they are not. Between the "useful arts" and
"technology" lay a world of difference.[10] The change in terminology flagged a profound transformation in American society in both material and linguistic terms
that came to be challenged along the way. Technology is a neologism. Only in the
1930s did technology become a keyword of American grammar and an all encompassing concept that could explain human life itself. It then came to denote the
useful application of scientific knowledge for the benefit of humankind, and engineers were designated as the sole bearers of that form of knowledge. Rather than a
neutral term, *technology* is itself part of a narrative production or plot of modernism, in which men are the protagonists and women have been denied their part.

   Technology presents us with an irony: the term *technology* could muster such
universal claims for itself in the twentieth century, but the modern meanings of
the term are of very recent vintage. Despite the evocative power of *technology* in the
late twentieth century, it failed to enjoy any currency in the nineteenth century.
First introduced by engineering advocates in the nineteenth century and theorized
by social scientists in the 1920s, the term gained wide and popular currency only
after the second World War. Yet the meaning of technology has been contested
and gendered throughout its history as women, workers, and African American
rights advocates sought to argue on the grounds of the international fairs, within
the halls of the patent office, and in the pages of dictionaries. As used by Raymond
Williams and applied in this book, a "keyword" locates both the descriptive and
prescriptive aspects in a defining process through which different groups – be they
industrialists, establishment engineers, social scientists, or women rights advocates
– helped shape meaning by contesting and contrasting these interpretations.
These historical actors take on rhetorical positions and use words that operate as
weapons in the contest with other. Words also produce metaphors – the very met-

aphors we live by – thus producing new meanings and experiences. More importantly, metaphors have the power to neglect and suppress information about human experiences of the world that does not fit the relation implied by the
metaphor. It is words, keywords, and metaphors that provide historians with windows through which to view a segment of the history of ideas and experience.[11] An
understanding of their workings is therefore essential for a history of gender.
Words matter. Technology is no exception.

We need to understand then that our use of the term as a keyword of American
culture is fundamentally new and to revisit its territory. The *Encyclopedia Britannica* included an entry on "Technology" for the first time only in 1978.[12] It asserted,
that technology is the exclusive knowledge domain of engineers, best embodied by
machines as the measures of men. The authors used technology to describe human
nature itself – the idea that a human being is essentially *Homo Faber*, a producer of
goods encompassing an entire system of people, means, processes, and artifacts.
"Technology is," the most revered encyclopedia concluded, "any means or activity
by which man seeks to change or manipulate his environment."[13] However appealing a metaphor, the notion of Man the Maker (*Homo Faber*) was a powerful intellectual construct that also had the power of neglecting the experiences and
material practices that did not fit the relation implied by the metaphor.

Most telling, the entry on *Technology* replaced and reworked an earlier entry on
Engineering Schools that described the emergence of American formal engineering
institutions like MIT, where its students engaged in banter to negotiate new
knowledge claims.[14] Most scholars have elected MIT as the lens through which to
view engineering education and the occupation at large, but to look at engineering
through that lens only runs the risk of myopia. In the country as a whole, schools
like MIT, the California Institute of Technology, and The Stevens Institute of
Technology were a minority within the ranks of technical education even though
they came to wield enormous ideological influence over the hodgepodge of other
engineering institutions. In the course of the nineteenth century, these institutions
began to claim a new kind of knowledge they called technology. *Britannica's*
renewed entry on *Technology* foremost obscured its origins from these academic
surroundings, enhanced the role of engineers in neutral sounding terms, and disguised contests over its meaning.

We therefore need to examine technology's transformation from an ill-defined,
little-used, and narrow concept to a keyword of American culture in the course of
the nineteenth and twentieth centuries; we need to rescue the original uses of the
word, follow its trajectory to understand why technology has become such a
deeply idiomatic and powerful expression in American culture that communicates
a gender-neutral set of meanings at the exclusion of others. To locate these strug-

gles and understand their outcome, we need to go back to the many communities that played a role in its formation.

Because the emergence of industrial capitalism involved both material realities and rhetorical strategies, I use a variety of sources ranging from engineering journals to dictionaries throughout this book. The chapters that follow show how the modern meanings are of recent vintage becoming widespread only after the second World War and after women and African-American rights groups had challenged its ascendancy. The embodiment of academic engineering knowledge incorporated machine aesthetics exclusively associating it with what Western men do, it also overlapped and competed with anthropological notions of material culture and civilization. Throughout the nineteenth century, women activists and women inventors' advocates like Gage and Charlotte Smith challenged the emergence of a new ideology that began to foreground establishment engineers and their corporate allies.

The second chapter takes a closer look at how the foot soldiers of industrial capitalism came to monopolize the term and practice of technology. It traces how notions of white manliness helped to revitalize class boundaries through three historical episodes when engineering was transformed from an elite profession to a mass occupation, and was remade into a middle-class occupation.

Engineering also helped build cultural infrastructure between 1890 and the 1930s. While at work on production floors, construction sites, and in the laboratories, engineers produced culture as well as goods. The construction of a male fraternal world in which women are merely bystanders is told through an exploration of the autobiographies of engineers at the turn of the century when America's imperial project came to a head, scores of younger engineers came to question the professional standards, and establishment engineers were looking for ways to gain cultural authority. The chapter shows how engineers of an older generation reworked a middle-class white man's world through the explicit exclusion of family and the people with whom they worked on a daily basis. Engineers spun their fraternal plots over the heads of (immigrant workers) as a play without women and non-western peoples. The issues of class identity shaped a male identity for engineers that focussed on technical details to the exclusion of workers at a historical juncture when engineering became deeply divided, segmented, and a mass middle-class occupation.

The next chapter further explores how this white, middle-class, and gendered male engineering identity was shaped in competition with female professional models during the decades of American overseas expansion in the 1890s. In the decades after the Chicago Columbian Exposition, popular Victorian writers like Richard Harding Davis and Rudyard Kipling brought this male white engineering identity into a broader cultural circulation. They cast it into a middle-class identity firmly linked to overseas expansion, aligned engineering and writing professionalism into a male alliance of sorts, and pitched it into sharp relief to Victorian womanhood. In their search for their own professional identity, male novelists magnified engineers as modern male heroes; but their late Victorian women colleagues like Mary Hallock Foote and Willa Cather questioned them. Earning a living with writing, they challenged the alliance between engineers and male authors who cast engineering as the antithesis of women's professional ideals and articulated an alternative language. Their articulation of a separate female culture both empowered them as professional writers and reinforced a separation between a male technical and a female artistic world.

A younger generation of modernist artists and writers who came of age just before the first World War began to caricature the sharply drawn Victorian divisions in male iconography of the technical world. With biting irony and exaggeration, Paul Haviland and others belonging to the modernist New York circle, used the sexualized machine metaphors as a means to bend Victorian notions of gender in search for modern models. Some early modernist women boldly appropriated and exploited the new male subject matter of machines and engineering. But in the cultural hierarchies, women artists who trespassed had to deal with different power structures than their male counterparts. In the end, modern artists also helped reinscribe rather than subvert the male iconography in technical objects through their graphic, often sexually and gendered explicit, language and images in the period between the world wars.

If the women writers, artists, and activists stood in a long tradition allowing them to articulate an alternative language, their slide-rule sisters within the engineering profession emulated rather than questioned male models of professionalism. The fifth chapter concentrates on the story of women engineering "firsts" whose history has been forgotten, erased, and reinvented in the last hundred years. It analyzes the kind of narrative devices available to women engineers, decodes their silence, explores why they became invisible both to themselves and to others, and how government propaganda and corporate practice helped cast them as bystanders to the technological enterprise.

As suggested by historian Mary Ritter Beard in the epilogue, however women have always been a force in, rather than bystanders, of history to quote her pioneering book of history. Beard, whose life goal was to show that history would be incomplete if women were to be left out, rescued them as inventors, engineers, and urban planners supervising sewer systems, designing houses, and sponsoring public services. She also demanded women's rightful place in the modern canon from archives to encyclopedias. But in the modern narrative productions of the twentieth century like encyclopedias, women were left out of the story. As told through the lemma of the *Encylopaedia Britannica*, technology is a narrative production and plot of our modern myth making.

Taken together, the chapters suggests why in our modern mythologies we consider middle-class white men like Morris Jr. as natural allies of the technical world that defies any explanation; they show how modern definitions of technology determine why we enter bridges under the definition of *technology*, but consider bras as outside its domain; they suggest why we believe that women when they enter the male constructed stage indeed looked like *deae ex machina* and why we continue to see them as suffering from the truest stage fright of all: technophobia. Foremost, these episodes show that throughout the last hundred years women have always been part of the cast of characters and have been engaged in an ongoing negotiation of their roles in this male play. We need to go back and review the history of engineering to re-examine what it meant not to be the son but the daughter of the occupation. The story of the term "technology" as used in the *Britannica* is both a curious and a crucial step in that reexamination.

I

Unsettled Discourses

Technology has been neither a keyword to American cultural grammar nor the exclusive preserve of engineers.[1] Language, quilts or corsets, all important objects of women's inventive activity in the nineteenth century, do not come readily to contemporary minds as significant inventions or as markers of technology today, yet they once were. An early nineteenth-century speaker could discuss manufacturing, industry, and industriousness, referring to any kind of production mechanical or otherwise that could even include agriculture; could mention science and useful knowledge in one breath without sensing any contradiction; could marvel about the wonderful inventions and discoveries that ran the whole gamut from languages to mechanical devices; and could speak of technology referring to academic knowledge as well as to the skills of millers, bakers, farmers, teachers, and innkeepers. This was a rich world, proudly displayed at world's fairs – the nineteenth-century carnivals of industrial life. Over the course of the century, different historical actors began to label, classify, and lay claims on these objects, activities, and knowledge domains, privileging some and discarding others; other lobby groups offered their own classification systems of knowledge and objects to counter this process. A century later, however, agricultural production, non-mechanical devices, languages, teachers, farmers, bonnets, and corsets were banished to the basements of the modern classification systems of technology. Then machines were put center stage as the measures of men and markers of modern manliness. These selections were hardly innocent choices, but the outcome of hard-fought battles. The history of the selection, labelling, and designation of objects as *technology* is essential for our current understanding of who is believed to be a true technologist or an inventor, who possesses the right kind of technical knowledge; and who or what may be the authentic bearer of technology. These struggles – conducted in both linguistic and material terms – went to the heart of the question of what constituted technology and what was to be excluded from it.

Long before engineering became a clearly defined profession, middle-class men began to stake out their claims to the myriad of activities associated with industrial production as a uniquely male prerogative. Regardless of its eventual outcome, the historical evolution of the word *technology* was neither straightforward nor self-evident. In fact, as a term technology was rarely used, if at all. Initially, in a rhetorical effort to establish their legitimacy, social groups ranging from industrialist boosters, leading scientists, engineering advocates, and public intellectuals to women's rights advocates, African-American educators, and anthropologists mobilized such terms as *useful knowledge, inventive genius, applied science,* and *the machine* rather than *technology* to claim their right and place in the polity. Among the great variety of groups anthropologists, academic engineers, advocates of machine aesthetics, and corporate spokesmen provided the terms that eventually would be incorporated into our modern neologism *technology.* By appropriating the idiom of science, industry, engineering, and anthropology these loosely defined associations bolstered a new male authority at the end of the century. The outcomes of these struggles have coalesced into our current understanding of what the term technology connotes, but only after much struggle during which the term shed many of its intimate associations with industrial labor only to become an emblem of Western man's superiority and civilization.

From the Useful Arts to Applied Science

The *Useful Arts* were a keyword to the American cultural grammar of the early nineteenth century. The founding generation of the Republic committed itself to the creation of a new American empire through economic growth. From Alexander Hamilton's *Report on Manufactures* (1791) to Thomas Jefferson's grudging realization that industry would have to join agriculture and commerce to safeguard the nation's liberty and prosperity, the so called useful arts were seen as the key to the Republic's internal, westward, and imperial programs. Manufacturing advocates began to employ the idea of the useful arts as an alternative to the aesthetic arts to bolster the emerging economic power of the bourgeoisie. Placing themselves in opposition to the aristocracy – or the 'parasites of wealth,' as they were occasionally called – advocates of manufacture argued that aristocratic forms of art were unproductive and worthless and served no purpose other than the aesthetic. In their discourse on the useful arts, manufacturers and their republican political colleagues borrowed from the Enlightenment philosophers Denis Diderot (1713-1784) and Jean Le Rond d'Alembert (1717-1783).[2] In North America, capitalist boosters did so by defining idleness and its opposite, industriousness or productivity, in monetary terms.[3] In these early discussions, in which manufacturers spoke of *manufacturing, industriousness,* and *industry,* this new branch of economic activity was not necessarily tied to machine production, but merely referred to a certain kind of production, which could even include agriculture.

In the early American Republic, the discussions on *idleness* and *industriousness* were conducted increasingly in gendered terms. The textile industry emerged as one of the first important manufacturing industries through which women and children entered the market economy as wage earners for the first time. Rapid economic changes during the 1830s began to mandate a reinterpretation of the young Republic's language of politics and gender. The textile industry became rapidly mechanized along with the important industries of flour grinding, saw milling, and iron production; it blossomed by embracing a system of bounties, patents, public funds, industrial espionage, organized emigration, trial-and-error practice, reading of technical tracts, informal talk, and illegal importation of machines and skilled workers from Britain. In a short period of time weaving workshops were transformed into massive textile factories where the many young farm women were drafted as wage laborers. In the early American Republic, the employment of women in the Lowells' cotton and Berkshire's paper mills was politically reformulated and justified. Not only did manufacturers argue that the employment of young farm women prevented men from being diverted from the agricultural sector, they also asserted that productive labor in textile factories rescued women from their inherent idleness. The heroic figure of the female wage earner in textile and paper mills sustained a unique American argument issued from the belief that the young Republic embodied a community of vigorous freeholders who had a civic stake in the polity. Female employment in the textile mills bolstered the idea that American industrialization could be different from its European counterparts. The healthy and upright farm daughters who moved from the family homestead to employment in textile or paper mills in New England would humanize and moralize a male body politics that exulted in electoral politics and rituals from which women were both implicitly and explicitly excluded. The rhetoric of American politicians and industrialists heralded female wage labor as a public boon, because women's work would preserve male work for the Republic's essential virtuous agricultural sector. Women and children were the *useful arts* incarnate.[4]

In these settings, the work of women and children operatives in factories also became semantically linked to machines – i.e. the ubiquitous industrial apparatus that in the discourses of early industrial capitalism received striking names like Spinning Jennies. In the imagination of factory owners, women and children could perform their labor in the same steady, predictable manner as machines that went through their repetitive mechanical operations, forever memorialized in Herman Melville's short story "Maidens of the Tartarus." The textile industry thus introduced the prism of gender by inserting women workers into the equation of men and machines.

A number of industrialists made common cause with the blossoming women's movement supporting women's education in the first half of the century. American industrialists, champions of the *useful arts*, sponsored educational institutions designed "to educate labor and set knowledge to work", welcoming the enrollment of farmers' daughters into the ranks of the necessary and disciplined work force that would feed the engines of American economic growth.[5] The idea of useful knowledge first identified with the useful application of knowledge fashioned after the formulations of Enlightenment philosophers, increasingly included the notion of profitability. Thus, the idea of *useful knowledge* or *useful arts* became identified not only with practical and useful application as the philosophers of the Enlightenment had formulated, but also with wages and commercial profits. In these settings, *useful knowledge* included needlework and metalwork as well as spinning and mining.[6]

The industrialists might have been eager to invest in establishments of learning that could set knowledge to work by constructing a notion of the innate utility of useful arts, but the emphasis on profitability of knowledge elicited a response. Earlier in the century *science* and *useful knowledge* were mentioned as if there were no contradiction. In the growing urban center of Philadelphia of the 1820s, managers established the Franklin Institute for the Promotion of Science and the Useful Arts modelled after the mechanics' institutes. Here as elsewhere science and useful knowledge operated side by side, but the Franklin Institute began to shed its broad commitment to technical knowledge for all designing separate lecture series for different groups: elevation for mechanics; instruction for youth; and rational amusement for women. A newly professionalizing group of pure-science advocates increasingly stressed the disinterested, not-for-profit motive in their search for knowledge. Science boosters like the President of the American Association for the Advancement of Science, Alexander Dallas Bache (1806-1867), or chemical scientist Ira Remsen (1846-1927) and physicist Henry Rowland, both at the Johns Hopkins University, invoked the growing authority of science in order to set themselves apart from the association of the *useful arts* with commercial enterprises. With increasing tenacity, *scientists* – also a new term in the 1830s – insisted on the insulation of science from profits; they depended for their professional identities on their ability to show the disinterested, pure, but nevertheless useful nature of their enterprise. In these struggles for professional recognition, Bache, Remsen, or Rowland's employment of the term *science* served as a rhetorical weapon against the perceived corrupting influence of those they feared most: the proponents of the useful arts. Their stress on impartiality and service, used as a counter argument to the useful-arts advocates' emphasis on the innate utility and profitability of knowledge, was also a matter of numbers; a growing army of practicing scientists and engineers working out in the field at the Erie canal, the industrial mechanics shops, and the railroad tracks began to outnumber scientists and physicists.[7]

Between these positions, a small band of engineering educators began to lay claim on another domain. Within the academic walls of some newly established engineering schools, technology emerged as a label of self-identification for a few ambitious engineering educators who self-consciously carved out a space between the useful arts and science during the second half of the nineteenth century.[8] The term *technology* was not exclusively reserved for engineers, but also sporadically used to include the kind of skills and procedures millers, bakers, farmers, teachers, and innkeepers needed in their occupations.[9] Early academic engineers reintroduced the term *technology* by extricating it from these artisanal associations and allying it to a more pristine scientific discourse. In this new sense, the term had first tentatively appeared in 1829 in the title of a series of lectures published as *Elements of Technology* by Harvard professor and physician Jacob Bigelow (1787-1879), who held a chair in the "application of science to the useful arts." By conjuring up the term, Bigelow sought to lay claim to a new domain of knowledge between science and the useful arts that straddled the genteel tradition of science and the plebeian ethos of hard work, endurance, and dirty fingernails on the shop floor, in the field, mine shafts, and engine rooms. Outside the small circle of academic engineers and students, however, few used the term "technology" to refer to a new form of knowledge or reality. Even few practicing engineers and scientists employed the term, if at all. Bigelow's term would be mostly forgotten for the rest of the century. It elicited so little response and recognition that even Bigelow dropped his own newly minted term only to revert back to a more current expression for the expanded version of his book ten years later, now called *The Useful Arts.* Indeed no other nineteenth-century lexicographers followed *Webster*'s cue on granting technology its own lemma for over a century.[10]

Industrialists, practical engineers, scientists, and engineering educators found the term *applied science* a much more powerful weapon for their rhetorical constructions than the term technology. Even so both terms, *applied science* and *technology*, ran far ahead of nineteenth-century material and social realities. While a few pioneering schools like Massachussets and California Institutes of Technology fostered the incorporation of scientific language in their search for legitimacy and financial resources during the final decades of the nineteenth century, the overwhelming majority of engineering institutions were unwilling or unable to spend significant funds on scientific research in order to buttress such grand claims even as late as the 1940s.[11]

Not everyone was allowed to employ these terms however little used or ill-defined. Northern and Southern industrialists considered the labor of African Americans useful in most social settings, but their skills, expertise, and experience were never to be labelled as *applied science*. In the Reconstruction era when a small band of Northern engineering educators began to claim applied science as their domain and expertise, African-Americans interested in technical fields were sent onto different educational paths. Between the Civil and the first World Wars, freed slaves were trained in the industrial arts at separate African-American man- ual training, industrial, normal schools, and land-grant institutions. At these sepa- rate institutions like Booker T. Washington's Tuskegee Normal and Industrial Institute in Alabama (1883) material practices might diverge from the rhetorical positions. The political realities of the American South prohibited Afri- can-American educators from claiming scientific labels, but they some devised novel strategies by teaching science subjects under other rubrics while carefully avoiding that contested discourse.[12]

The famous controversy between the Southern ex-slave Booker T. Washington and the Harvard educated Bostonian W.E.B. Du Bois (1868-1963) centered on what "useful" meant to African Americans and focussed on the issue of Afri- can-American vocational and technical education. Supported by Northern busi- nessmen-philanthropists and Southern whites, Washington believed that technical education (useful knowledge) would turn African Americans into self-sufficient workers and dignified first-class citizens, but Du Bois radically op- posed the notion that knowledge should always be useful, profitable, and commodified. In the light of their slave labor past and their economic exploitation, the greatest liberation of African Americans and the finest marker of their manli- ness was the right not to be useful, Du Bois believed. He preferred the right to be genteel, intellectual, and professional in the cultural world of Shakespeare and Soul, over joining the ranks of skilled labor in the employ of industrial capitalists.[13]

Thus communities ranging from industrialists to African-American educators began to lay claim on the late nineteenth-century world. The world of the useful arts from antebellum to Gilded-Age America enclosed different constituencies that converged and overlapped, but also contradicted, clashed, and contested each other. Industrial advocates, philosophers, engineering educators, Afri- can-American leaders, and practical and academic scientists gave meaning to their experiences and hopes through the employment of terms like *useful arts, utility, ap- plied science*, and *technology* that operated as weapons; they articulated intellectual constructs and material practices. Words and concepts like technology fell out of favor while others were readily used. In their disputes and communications, these historical actors also created new shades of meaning, embracing some historical ac- tors, and excluding others. The nineteenth-century international fairs became the prime sites for this selection process that eventually would separate the alphabet, corset, and bonnets from steam engines, trains, planes, and cars as the true, gendered, and racialized objects of technology. The nineteenth-century world's fairs became an important podium on which *the useful arts* were staged.

In the years before motion pictures, radio, and television, nineteenth-century contemporaries went to see worlds' fairs to sample and experience the world. Praised as *world universities* and *workshops of the world* by their boosters, world's fairs were the *encyclopedias of civilization*, *rituals of display*, and *competition between nations* of the nineteenth century. Fair-goers from all social classes visited them to be amused, instructed, and diverted from the industrial sorrows that were raging around them. Before the expansion of such forms of communication as technical magazines, catalogues, advertising, and professional engineering conventions, world's fairs also provided manufacturers, new professionals, and activists with a primary forum, meeting point, and international network. It was the place where manufacturers and engineers sold their goods and ideas, disseminated information about new products and scientific discoveries, and bolstered their knowledge claims on the nineteenth-century world.[14] Many other nineteenth-century aspiring professionals – including librarians, historians, scientists, and engineers – gathered there to launch their professional organizations to lay their own claims on the world around them.

In the U.S., the Smithsonian Institution supervised most fairs and provided the intellectual frameworks for them. Organizers did not merely put dresses, bonnets, corsets, books, reapers, and steam engines on display, but also helped devise systems that became instrumental in classifying both knowledge and objects for the nineteenth century. No wonder fair grounds were contested terrains. Manufacturers sought to promote their products, nations competed for prominence and investments, and different lobby groups – ranging from engineering advocates to women rights groups, African American activists, and labor leaders – fought for the inclusion of their causes into these encyclopedia's of civilization. The public exposure of these fairs was as large and effective as any lobby group could hope for: between 1876 and 1916 nearly one hundred million people visited 12 world's fairs throughout the U.S. Women and African-American rights advocates eagerly seized these occasions to challenge the new juncture of civilization, inventions, and white manliness as they would demonstrate in the Eastern industrial hubs of New York in 1853 and Philadelphia in 1876, the Western gateway in Chicago in 1893, or the cotton capital of Atlanta of the New South. Fairs were the true staging ground where conflicting understandings of technology were put on display. From 1876 the Smithsonian Institution provided the crucial support, personnel, and materials for the ethnological displays from its ethnological and anthropological departments. Most importantly it proved to be instrumental in helping to layout the exhibits in a classification system of the nineteenth-century world that included women and non-western people. Significantly, after the first World War the National Research Council (NRC) and the Departments of Commerce both closely connected to the corporate-military complex would take on that role for the twentieth century.[15] This change would be as significant as far reaching.

FEMALE FABRICS VERSUS MANLY MACHINES

Well into the twentieth century, *inventive genius* was not necessarily understood to be machine-bound. Inventions included the entire gamut that ran from fabrics, language, arts, and mythology to mechanical devices. In the decade leading up to the 1876 fair in Philadelphia, definitions of innovation and human ingenuity were still in flux and subject to negotiation, as the work of the cultural anthropologist Lewis Morgan and women's rights activists shows.

Social anthropologists, among whom Smithsonian scientists played major roles, lent stature to theories on the course of civilization and the primary defining role of inventions. In mid-century, social anthropologists began to mobilize the idea of inventive genius as an index of civilization. This measure, peripheral at first, found its way into many a social anthropologist's overarching theory of the development of civilization, and eventually turned into the yardstick of a nation's overall progress. Lewis Morgan (1818-1881), the patent lawyer, businessman, railroad investor, and social anthropologist, accorded the notion of inventiveness a special place in human evolution theories in his *Ancient Society* published a year after the Philadelphia Centennial exposition in 1877. He considered the rate of inventions as the prime mover in the evolution of societies, pointing to "inventions and discoveries" as the keys to society's ability to move up or down the evolutionary scale. In the first part of his thesis, entitled "Growth of Intelligence through Inventions and Discoveries," he asserted that "the most advanced portions of the human race were halted...until some great invention or discovery, like the domestication of animals or the smelting of iron ore, gave a new and powerful impulse forward."[16]

Morgan's overarching theories on inventions included women because he incorporated the nineteenth-century nonmechanistic interpretation of inventions. This inclusion occurred in an accidental manner but proved to be nevertheless important. After all, Morgan's purpose was to validate a hotly debated proposition that absorbed the attention of intellectuals in both Europe and the U.S. in the era after Charles Darwin's publication of the *Origin of Species* in 1859: all human races shared a common origin and were monogenetic rather than polygenetic. Because the study of non-Western cultures served as a time machine for Morgan and his colleagues, one that allowed them to look at the origins of human evolution, he sought to establish linkages in kinship systems, customs, and cultural attitudes among the Iroquois and other American Indian peoples. Despite his practice as a patent lawyer for the railroads, Morgan did not limit his understanding of inventions exclusively to patent activity because of his focus on domestic institutions – a natural outgrowth of his earlier work on the structure of family ties. He included in his genealogy of inventions artifacts handled by men, like bows and arrows, but also female skills like basket weaving. By juxtaposing inventiveness and domestic institutions – at least for other cultures – Morgan unintentionally presented women's daily work as an useful activity on a par with men's. He suggests that women, too, were inventors.

Early in the century, inventions still embraced an array of human products from intellectual to practical skills, from corsets to cutting machines, but in the 1890s true and important inventions increasingly took on the form of machine-bound and patented objects. This change of emphasis would eclipse the nonmechanistic and nonpatented formulations of inventions then fashionable in the field of anthropology; it also would eclipse the role that social anthropologists had initially accorded to women as inventors of nonmechanistic objects. The place that inventions came to occupy in the emerging understanding of technology became so central that it served as a stock political argument and benchmark for gender and racial differentiation. Over the next century, much was at stake in the question of which nation state, social community, or racial group could lay claim to the highest rate of inventions. In the emerging paradigm, the answer to the rhetorical question of whether or not women possessed inventive genius was thought to bear on the issue of women's worthiness as full participants in the body politic. Because of the enormous political weight accorded to them, inventions were a hotly contested terrain throughout the century, and the scrutiny of women's inventive capacity was no exception to the national pastime of counting patents.[17]

The changing meanings of these terms categorizing inventions did not occur without the intellectual intervention of women activists. In mid century women's rights advocates questioned the appropriation of inventive ability as a male preserve. They challenged the new idea that inventiveness represented an exclusively male prerogative. More fundamentally, early feminists positioned women at "the handle of the crank": women were the catalysts who pushed human evolution upward and onward to the next stage of progress.[18] Several feminist critics went so far as to question the premises of male inventiveness altogether. Women activists targeted an array of male institutions that increasingly canonized technology as a male preserve; they voiced their dissenting opinions again and again on the grounds of the World's Fairs in 1853, 1876, and 1892 and in the halls of the Patent Office on the occasion of its Centennial Celebration in 1890. They would contest the nineteenth century's inventories of the world such as archives, dictionaries, and encyclopedias. These occasions also served as rallying points for feminists to

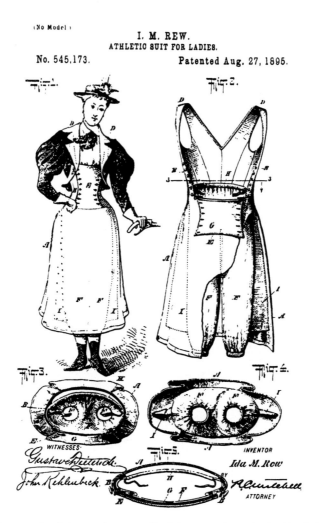

(No Model)

I. M. REW.
ATHLETIC SUIT FOR LADIES.

No. 545,173.   Patented Aug. 27, 1895.

WITNESSES   INVENTOR
Gustave Dietrich   Ida M. Rew
John Kehlenbeck   BY R. Cumberland
                                              ATTORNEY

Figure 1.   An example of nineteenth-century feminine ingenuity authorized by the U.S. Patent Office. Ida Rew's 1895 athletic suit engineered a balance between freedom of movement and sense of propriety for middle-class women.

forge an alliance with the hundreds of female inventors – working wives and widows like Martha Coston, Harriet Hosmer, Helen Blanchard, Josephine Cochrane, and Ida Rew who worked to reap the profits of their ideas – and to advance their broader political agenda.[Figure 1] Women intellectuals from Matilda Joslyn Gage, Ida Tarbell, Charlotte Smith, Minnie Reynolds to Mary R. Beard championed female inventive activity in the post-Civil War period. Their strategies ranged from lobbying for women's equal representation to confronting the exclusionary policies and the building of alternative but often segregated institutions – whether the establishment of separate pavilions, the mounting of fairs, or

Figure 2.   Portrait of Joslyn Gage, theoretician of the radical wing of the women's movement, who criticized the newly sanctioned centrality of patented inventions and demanded equitable distribution of resources for men and women. Here portrayed in the pensive pose of an intellectual. Courtesy of Schlesinger Library, Radcliffe College, Cambridge, MA.

the composition of women's encyclopedias and biographical dictionaries entirely devoted to women's contributions to civilization. They countered the emerging male genealogy of inventions that sought to prove that only men had the intellectual acumen to produce significant inventions.[19]

Women rights advocate Matilda E. Joslyn Gage (1826-1898) was the first American woman to challenge the male genealogy of inventions and to formulate the clearest ideological statement on female inventiveness as early as 1870.[Figure 2]

Together with Elizabeth Cady Stanton, Gage was the intellectual force behind the radical wing of the women's rights movement. Active in the women's rights campaign since the 1850s, she held the position of president of the National Woman Suffrage Association, from which she resigned in May 1876, just before the opening of the Philadelphia Centennial Exhibition, in order to give center stage to the more famous Stanton. Years before the Philadelphia Exhibition of 1876 and Morgan's ruminations on the importance of inventions as an index to the ontogenetic stages of civilization, Gage had argued in a 1870 suffragist pamphlet that women inventors did indeed exist. In this pamphlet, entitled "Woman as inventor," she had rescued Catherine Littlefield, the widow of General Greene, from oblivion by describing her as the principal mind and financial sponsor behind the cotton gin patented under the name of Eli Whitney. Gage's argument was quite ingenious because she positioned women at the center of both national economic development and the invention of silk production. Engaging emerging nationalistic arguments, Gage maintained that "these two inventions by women of silk and the cotton-gin have done much to build up the State, to define social and political positions and to further the interests of mankind." She challenged the individualist understanding of inventions by arguing, first, that progress was a result of small and incremental changes: "All progress in the arts, in science, in wisdom, is the result of successive steps; and it is impossible to foresee the consequences which may arise from the omission of an act by even the most obscure person."[20] Second, she pointed out that many inventions could not be traced to a single author or era. Third, she contested the patent system and the manufacturing industry as society's exclusive sources for understanding inventive behavior, since "the patentee is not always the inventor; neither is the manufacturer always the originator." Her critique centered in part on the narrow interpretation of the inventive process. To consider patents the only tangible evidence of inventive behavior was incorrect.

Gage based her argument on the same – and at the time widely accepted – taxonomy of inventions as Morgan's. In her 1870 treatise, she included a range of human products she considered important inventions: mechanical devices, fabrics, language, arts, and mythology. Thirteen years later, however, when she wrote a second version of her treatise on women inventors, the crux of her argument had shifted, reflecting the emerging understanding of human inventiveness as being inextricably linked to the search for bigger and better machines. In 1883, her definition of inventiveness was confined to things mechanical. Writing for a national audience in the well-respected *North American Review*, Gage set out to show that women were not only genuine inventors but that they possessed "mechanical genius" on a par with men.[21] Her article marginalized such nonindustrial or non-object-related fields as literature, arts, and language. The connection she now drew between inventiveness and political freedom formed the crucible of her argument:

"The inventions of a nation," she argued, "are closely connected with the freedom of its people," and omitting women from the inventive process would endanger the progress of human evolution. At this point echoing Morgan, she concluded: "No less is the darkness of the world kept more dense, and its civilization retarded, by all forms of thought, customs of society, or systems of law which prevent the full development and exercise of woman's inventive powers." It mattered little to her whether or not women possessed *mechanical genius* – what mattered more was an equitable distribution of, and access to, society's economic resources. Given that women lacked the most basic economic and political rights, Gage argued, she thought it remarkable that women should be capable of inventing anything at all.

In her 1883 treatise, Matilda Gage tacitly acknowledged and engaged in the paradigmatic shift towards a machine-bound interpretation of inventions. But these more explicitly gendered definitions of human creativity proved problematic for women, since the tendency to glorify machines was associated increasingly with new developments in the machine-tool and steel industries. Few women worked on these production floors, as reformer journalist and business historian Ida Tarbell would point out in her article on women inventors. In particular, the steel industry's capital-intensive apparatus like Bessemer converters from the hills of Pennsylvania to the city limits of Chicago functioned as a symbol of power – one might label it a fetish – among capitalist entrepreneurs.

Thus Gage's parameters of the debate concerning human ingenuity shifted dramatically in the period between 1872 and 1883. In just ten years Gage moved from viewing human inventiveness as an gender-neutral activity to a talent that was the prerogative of men – from mere discoveries to machines, from genderless activities to male marked objects. Her shift in focus reflected an array of actual social changes and class tensions in American society. On the heels of major labor disputes in Pennsylvania and a national economic panic, the 1876 Centennial celebration at Fairmont Park in Philadelphia, with the Machinery Hall and its Corliss engine on center stage, both visually and viscerally foregrounded and forged the machine-bound interpretation of inventions. The machine, embodied in the smooth-running and humming Corliss' engine, became a national icon marked as male and middle-class. Characterized as "an athlete of steel and iron," the Corliss engine appropriated a sense of national unity that belied the major divisions between capital and labor, between men and women, and between white and black in American society. The allure of machines would continue to increase in the twentieth century, when technological sophistication became a trope designed to authenticate male authority in American society and to corroborate the inherent superiority of the Western world." Soon machines would mark male middle-class power in the West.

The prominent display of giant engines and massive machines in Philadelphia elicited negative commentary from women's groups, however, under the leadership of the conservative woman activist Elizabeth D. Gillespie, the granddaughter of Benjamin Franklin, various women's organizations raised enough funds to construct a Woman's Pavilion at the Philadelphia Fair, to "give to the mass of women who were laboring by the needle and obtaining only a scanty subsistence, the opportunity to see what women were capable of attaining unto in other and higher branches of industry."[23] [Figure 3] Although women contributed and won prizes throughout the Fair – in the fine arts, education, and farming, for example – the point of the Woman's Pavilion was to highlight women's share in industrial life. To lend greater authority to the Centennial sisterhood, Gillespie devoted special attention to attracting women inventors. In an effort to destabilize that other "humming" metaphor of industrial life – the giant twin-cylinder Corliss steam engine towering over all other equipment on display in the Machinery Hall – the organizers appointed a female engineer named Emma Allison as "the presiding genius of the engine room" in the Woman's Pavilion. In this separate women's domain Allison became a beloved attraction by operating a portable Baxter steam engine that drove six power looms at which women wove carpets, webbing – and silks. She thought herself quite capable of running these machines and declared that it was far easier for her to operate the steam engine than it was for mothers to "operate" their children. Allison's steam engine supplied sufficient energy for a carpet loom, a quill wheel and a spooling machine, a ribbon loom, a Jacquard loom, as well as the cylinder press which printed *The New Century for Women* for the duration of the Fair. The world of the Lowell textile mills had been a female arena for decades, but the twin-cylinder Corliss steam engine so dominated the public's imagination at the Philadelphia Fair that a woman reporter of *The New Century for Women* criticized visitors who watched the Corliss engine in awe but failed to admire the intricate machinery operated in the textile mills by women.[24] Her remark lay the foundation of a series of competing images as the true objects of technology that would linger on throughout the century: female fabrics and male machines.[Figure 4]

Women activists were hardly united on the issue of female participation in the fair. In the opinion of radical women's rights advocates like Matilda Gage, Susan B. Anthony, and Cady Stanton, however, Gillespie's efforts to parade women's work and their inventiveness were useless. They argued that the Woman's Pavilion and the female presence throughout the exposition, calling attention to the accomplishments of women's inventors, failed to disguise the grim fact that women were still denied the right to vote.[25] Gage and other women suffragists dismissed the Woman's Pavilion because it did nothing to reveal a "true exhibit of woman's work." Most of the work done by women, they thought, took place in a business

Figure 3.   Women's Centennial Executive Committee planning to parade women's skills, products, and inventions to demonstrate women's equal worth in civilization at the Philadelphia Centennial Exposition of 1876. These conservative women were opposed by radical activists, who protested against women's legal and political inequality throughout the fair. Reproduced from *Frank Leslie's Illustrated Journal Historical Register* of 1876.

environment they neither owned nor controlled. According to the most vociferous champions of women's rights, no matter how many women inventors might have been present at the exposition – there were about seventy women who were demonstrating their talents to the public – "the most fitting contributions to the centennial exposition would have been these protests, laws and decisions which show [women's] political slavery."[26] If married women were not even granted the right to control their own earnings, as Gage wrote later, "should such a woman be successful in obtaining a patent...Would she be free to do as she pleased with it? Not at all... She would possess no legal right to contract or to license any one to use her invention."[27] While Gage criticized the newly sanctioned centrality of (patented) inventions, she nonetheless conceded that the patent system was an important site of inventive behavior; Gage, in other words, altered her views without giving up on her desire to confront the dominion of men in the field of mechanical ingenuity.

Figure 4.   Corliss's giant steam engine located in the middle of Machinery Hall - here set in motion by President Grant and Emperor Dom Pedro II to open the Philadelphia Centennial Exposition on May 10, 1876 – came to symbolize America's manly industrial capitalism. Called "an athlete of steel and iron," its size, status, and position competed with the less visually spectacular women's skills and products on display at the Woman's Pavilion. Reproduced from *Frank Leslie's Illustrated Journal Historical Register* of 1876.

Figure 5. Business historian, muckraker, and moderate women's activist, Ida Tarbell (1857-1944) at work in 1917. She argued that women's needle and men's machine work were equally valuable creations. Permission and Courtesy of Culver Pictures, New York, NY.

The journalist, business historian, and lecturer Ida M. Tarbell (1857-1944), best known later in life as a muckraker against Standard Oil chose a different strategy when she questioned the male genealogy of inventions. [Figure 5] In an effort to subvert the newly dominant paradigm, the then thirty-year-old Tarbell who still felt strong affinities for the women's movement at that time and worked on the staff of the *Chautauquan* magazine in the hills of Pennsylvania, tried to elevate the stature of women's domestic inventions in the hope they would no longer be dismissed as insignificant and inconsequential. Instead of trying to beat men on their own turf, she reclaimed the older meanings of inventions. She placed women's inventive creativity firmly in the framework of the separate-spheres ideology that had become a stock argument of the middle class and her circle of educated women. In her 1887 article for the *Chautauquan*, a magazine that offered adult education to traditional women who had come to subscribe to the cultural premises of separate spheres, she pointed out that women did not work in machine-tool related industries.[28] She argued that because "it was reasonable to expect that ingenuity will be exercised proportionately to opportunity," one had to look elsewhere. Tarbell then rescued women's practical solutions at home as legitimate inventions because they were effective and valuable, thereby undermining the new discourse concerning women's supposed idleness. "An invention is an invention whether it be for house work or mill-work, and the kind of mental quality it requires is the same," she contended. Much creativity and skill were required for mothers to alter their daughters' dresses and then use the same fabric for bed quilts. "It is the habit to speak of these women as 'handy' or 'full of ideas,'" but she insisted that "such women are inventors; their work, inventions." In spite of her recapitulation of the older meanings of inventions in her own reformulation, Tarbell could not escape the emerging machine-bound understanding of inventions because, like Gage, she felt compelled to argue that women could indeed be successful in devising things mechanical.

At the end of the decade, Charlotte Smith (1843-1917) followed in Gage's footsteps and took up the cause of women inventors by entering the hallowed halls of the newly constructed temple of inventions, the U.S. Patent Office in the nation's capital. She championed working-class women rather than Tarbell's middle-class women who worked at home. Smith, president of the Woman's National Industrial League of America, directly challenged the Patent Office in a blitz campaign on the occasion of its hundredth anniversary in 1890. Unwittingly, her challenge only reinforced the notion that genuine inventive activity was to be found in the patent business, lending it a kind of authority that Tarbell had denied and Gage had questioned. A flamboyant campaigner for working women and a shrewd congressional lobbyist, Smith defended women inventors because she recognized that inventions could be a significant source of income for women. She published *The Woman Inventor*, a magazine designed to coincide with the Centennial celebration for the Patent Office. [Figure 6] In preparation for the anniversary and her own campaign, she first managed to persuade the Patent Office in Washington to compile a list of all women inventors since 1790.[29] But as historian Autumn Stanley has documented in her study of female patentees, when the Patent Office's well-intentioned clerks compiled a list for Smith in 1892 they glossed over women's mechanical inventions – a further indication of the growing importance attributed to machinery and women's exclusion from its domain. By reexamining the patents issued for 1876, when the Gillespie-sponsored women inventors exhibited in the Woman's Pavilion at the Philadelphia fair, Stanley found that the compilers had omitted one woman's invention for every four they recorded; she also concluded that these omissions on the list generated for Charlotte Smith were not random. Machines represented the largest single category of the omitted inventions, compared with categories like agriculture, chemistry, furnishings, health/medicine, heating, cooling, domestic labor-saving devices, and clothing. Moreover, Stanley found that the mechanical devices the patent clerks omitted from the list were "strikingly nondomestic or what might be called nontraditional inventions for women."[30] Thus, despite Smith's feminist intervention and the help of cooperative

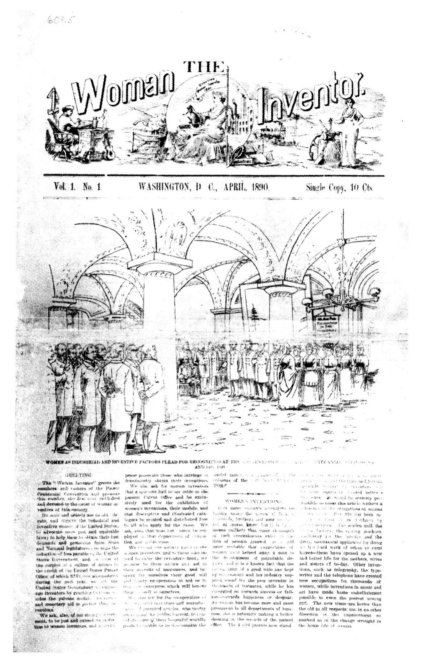

Figure 6.   Etching in *The Woman Inventor* dramatizing a group of women inventors who petitioned for recognition and fair treatment in the newly constructed Hall of the Patent Office during its centennial celebration in April 1890. Courtesy of Archives Center, National Museum of American History, Smithsonian Institution, Washington, DC.

clerks, the late nineteenth-century Patent Office's list of women inventors provided a fascinating example of how new mentalities obscured the vision of federal
employees in Washington DC, who appeared unable or unwilling to even *see* the
tangible results of women's "mechanical genius." Such a suppression reflected a
paradigmatic shift in American perceptions of the nature of inventiveness as machine-bound and masculine.

In her challenge to the masculine and middle-class taxonomy of inventions,
Charlotte Smith did not limit her efforts to the national preoccupation of counting patents. In her call for action, Smith went even further in reclaiming inventions for women by demanding that the Federal Government protect women
inventors and prosecute "those who infringe or fraudulently obtain their inventions." She argued that the Patent Office should establish a special office for
women which would display their inventions and label them properly. She also
called for solidarity among women inventors: "We have the genius, but it requires
development and encouragement, hence let us...encourag[e] one another."[31]

Otis T. Mason (1838-1908), the Smithsonian Institution's anthropologist, curator, and classifier, answered Smith's bold challenge to the male institutionalization
of patents. As chief curator of the Smithsonian's Department of Ethnology, Mason was principally responsible for popularizing the evolutionary scheme of human development and the scientific racism that went along with it at the American
International Fairs in the nineteenth century. He first visualized his view of
women's industrial role in the grand scheme of evolution in his ethnographic display starting with the 1893 Chicago Fair exhibit "Woman's Work in Savagery" he
installed for the Woman's Pavilion at the request of women activists. Mason, a disciple of Lewis Morgan and influenced by German thinkers, had a vested interest in
placing inventions in an evolutionary rank-order. He also regarded the frequency
and innovative quality of human inventions as an inventory of social progress
which served, therefore, as an essential key that might be able to unlock the mys-
tery of differential evolutionary paths on the basis of cross-cultural comparisons.[32]
Filling in the broad outlines Morgan had provided, he felt a special intellectual ur-
gency to examine the role of female creativity. Mason explicated his overall theories in his writings on women's inventiveness, and he popularized and visualized
them through his design of the ethnographic displays at World's Fairs – some at
the request of women's champions. He also endowed them with further public
stature in his centennial address at the Patent Office. In his speech on the occasion
of the hundredth anniversary of the Patent Office in 1890, Mason argued for a
much longer genealogy of inventions than the advocates of industrial capitalism,
who had gathered to "glorify the nineteenth century." Both men and women, he
emphasized, had always functioned as inventors. He posed a rhetorical question:
who "quarried the clay, manipulated it, constructed and decorated the ware,

burned it in a rude furnace and wore it out in a hundred uses?" To an audience of manufacturers, politicians, government officials, and engineering educators who had assembled for the Patent Office centennial occasion he argued: "Over and over again, those who have preceded me on this platform have pointed to James Watt as the true deliverer of mankind. Far be it from me to take one leaf from his laurel crown; but the inventor of the alphabet, of the decimal system of notation, or representative government, of the golden rule in morality, were greater than he." While Smith lobbied to protect women inventors, Mason pondered the meaning of women's real patent: "The best woman to cook or sew or carry loads got the best husband. That was her patent."[33]

In the halls of the Patent Office, Otis Mason continued on the path charted by Lewis Morgan, but he also kept a safe distance from Smith's working-class women inventors. He promoted the evolutionary and comparative approach: at the Southern Expositions in Atlanta and Nashville a few years later, he mounted life-size replicas of Native American, African, and Polynesian women performing productive work, as he labeled it. He showed women of other cultures weaving baskets and netting, and he held them accountable for humankind's inventions because "Women, among all the primitive peoples, were the originators of most of the industrial arts."[34] Mason read in his objects – tools, artifacts, and skeletons – "the stories of their owners many centuries ago" and concluded that most of the artifacts he examined had been invented and used by women. He argued his case in *Woman's Share in Primitive Culture* (1894). In excerpted and popularized form it also appeared as "Woman as an Inventor and Manufacturer" in *Popular Science Magazine*, a journal that served as an important forum for debates on the social and political consequences of gender differences. Mason used Lewis Morgan's notion that the rate of inventions is an index for human progress to suggest that woman, as food-bringer, represented "the earliest inventor" and that her "ingenuity has been an important element of progress" since the early stages of human evolution.[35] Like Morgan, Mason considered food preparation, weaving, art, and language to be important discoveries and inventions.

Mason devoted much time and attention to highlighting women's economic, industrial, and inventive activities in his all-embracing theory, for which he would be gratefully quoted by feminists, but he viewed women's skills as atavistically belonging to an earlier era. By equating working-class women with women of "dusky" and "savage" cultures, Mason racialized working women.[36] The comparison went beyond the analogy; he considered working women as actual remnants of an archaic, less civilized age of the past. To Mason, in other words, wage-earning women in America's urban neighborhoods resembled a primitive tribe in the midst of civilized middle-class America. In the end, he dismissed women like Charlotte Smith and her working-class proteges as well as female inventors as irrelevant and inconsequential in the larger scheme of human evolution. The cross-cultural representations at the fairs reinforced the notion that middle-class women's leisure time in American society signified progress. Machine-bound technology remained thus safely inscribed as male and Western.

In the years leading to the 1893 Columbian Exposition, prominent women including Susan B. Anthony had done everything in their power to question the course of civilization as an exclusively male endeavor which Mason and other Victorian anthropologists were promoting. Early on, they pressed the fair organizers to place placards informing fair-goers what proportion of each exhibit was produced by woman's labor, urging women to submit special exhibits throughout the fair, and trying to make sure these submissions received equal treatment in the selection process. Their desire to show that women were an integral part of civilization met with complete resistance. As a last resort, they settled for a separate and segregated building, "The Woman's Pavilion," designed by the young MIT architecture graduate Sophia G. Hayden under the auspices of the officially sanctioned Lady Board of Managers headed by Chicagoan socialite and feminist Bertha Honoré Palmer (1849-1918). Choosing from among 3000 patents submitted by women and available on file at the Patent Office, the Lady Managers arranged the display of what they considered truly useful inventions in an Invention Room where Olivia Flynt demonstrated her health corset, Josephine Cochrane her dishwasher, and Martha Coston her night signalling system adopted by the Navy.[17] [Figure 7] Reflecting women's small space of negotiation, Hayden's pavilion was the smallest of all and precariously located between the official White City and licentious Midway Plaisance, between white manliness and the dark effeminate races, between the manly Court of Honor – celebrating the seven virtues of civilization through Manufactures, Mines, Agriculture, Art, Administration, Machinery, and Electricity – and the effeminate uncivilized, barbarous dark races. Even if the Lady Managers questioned the linkage between manliness and civilization, they did not dare to upset the racial hierarchy in the Columbian "exhibition of the progress of civilization in the New World." They made common cause with the racial taxonomy through their sponsorship of Mason's ethnic display and their refusal to answer calls from leading African Americans including journalist Ida B. Wells and Frederick Douglass for inclusion.[18] In the end, the machinery of the brave new world was not only safely inscribed as male, middle class, and Western, but white as well.

Following the decades of the Columbian Exposition, the modern art movement exploded but ultimately reinforced the male, white notions of machines in their celebration of the machine aesthetics. Their evocation of the machine in word and image turned into a powerful pillar for the modern understanding of technology. In the early teens, the machine became a buzzword of modernism

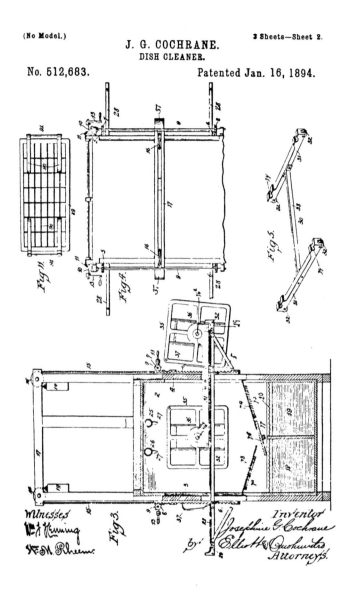

Figure 7. Patent drawing of Josephine G. Cochrane's dish washer exhibited in the Woman's Building Invention Room and used in most large restaurants at the Chicago World's Fair of 1893.

with explicit male iconography. American artists – among whom the Precisionists, Futurists, and Dadaists most graphically – explored machines as male models, metaphors, and microcosms of modern life. Late Victorian writers and a new generation of modernists mobilized older transcendentalist writers like Carlyle for a new understanding of machines that was understood to be at once material and metaphorical. This linkage had not been expressed so clearly up to this point. The metaphorical and visual allure of machines continued to be immensely appealing, precisely because of their concreteness and materiality. To the modern world, they also turned into fetishized objects of technology. As the world's fairs had done, they became the materialized gauges of Western identity, superiority, and hegemony. As such, machines functioned as measuring devices by which Western cultures gauged themselves with increasing confidence and assessed other cultures with increasing condescension. The selection, labelling, and designation of objects as technological proved essential for a new understanding of what constituted technology, and what did not. Language, quilts, or corsets, all important objects of women's inventive activity throughout the nineteenth century, were increasingly filtered out of contemporary minds as significant inventions or as markers of true technology. Under the auspices of a generation of cultural anthropologists, the American Smithsonian Institution had proved to be instrumental in helping to layout the exhibits in a classification of the nineteenth-century world that could still include women and non-Western people, but after the first World War this would become less viable.

VEBLEN AMALGAMATING, ENGINEERS, MACHINES, AND TECHNOLOGY

The events of the first World War and the modern arts discourse mobilized the once discrete and dispersed discourses of the useful arts, applied sciences, technology, inventive genius, and machines into the more unified one we now call technology. By the 1920s technology was conceived as a self-contained, self-generative, and machine-bound object devoid of human agency, and engineers came to play a prominent role in it; they were thought to devise and supervise it. This remarkable and dramatic twentieth-century revision of technology had little to do with Bigelow's usage or even with engineers' employment of the term. As late as 1932, the public intellectual, political scientist, and historian Charles Beard (1874-1948) could still believe that the term technology "is freely employed in current writings [but] its meaning as actuality and potentiality has never been explored and defined."[39] Once the term had moved into Beard's Progressive intellectual circles through discussions about the notion of technological unemployment, it migrated out in the late 1930s to become rather clichéd. As a keyword in a new cultural grammar, it first appears in the work of the institutional economist Thorstein B. Veblen (1857-1929). Because of his status as major public intellectual at the turn of the century, we can trace the transformation from the Victorian to the modernist cultural grammar through his work. While for Gage, Morgan, Tarbell, and Mason it had once been possible to conceive of women as active agents, inventors, and workers in the development of human evolution, Veblen further privileged male engineers, and excised women and workers – a rhetorical strategy later canonized by *The Encyclopedia Britannica's* lexicographers. It is through his work that we can carefully trace the migration of the term and witness its paradigmatic and semantic shift, which makes his oevre worth a careful reading. Veblen not only helped to revive the term *technology*, but also merged the three divergent nineteenth-century rhetorical positions into one: male machine metaphors, engineering professionalism, and cultural anthropology.

Trained as an economist and anthropologist, Veblen was acclaimed by many disciplines as their paragon. When in 1938 *The New Republic* asked leading intellectuals to name the books that had shaped their minds, the marginal academic Veblen was first on the list. During the late 1910s and 1920s, "everyone of intellectual pretensions read his works," as the conservative cultural critic H.L. Mencken remarked. "There were Veblenists, Veblen clubs, Veblen remedies for all the sorrows of the world."[40] Roosevelt's intellectual spokesmen (e.g. Rexford Tugwell and Felix Frankfurter), Veblen's colleagues at the New School (e.g. Charles Beard and Wesley Mitchell), leading left-wing publicists (e.g. Stuart Chase and Max Lerner), social scientists (e.g. Robert Lynd and William Ogburn), and advocates of Technocracy all considered his work brilliant and seminal. Veblen crossed many disciplinary boundaries, profoundly shaping the public discourse through his own work and his disciples in the period between the world wars.[41] By the 1940s, female fabrics and the useful arts were no longer viable in the cultural grammar of the U.S.

Thorstein Veblen was above all a wordsmith, a master crafter of the English language, and an inventor of words, in short, a coiner of keywords for which he became famous. Always witty, ironic, and biting, he focused on the changing meanings of words and would analyze them, turning commonplace terms upside down. In a memorable character sketch, the writer John Dos Passos described Veblen as "a man without smalltalk... [whose] ...language was a mixture of old mechanics' terms, scientific latinity, slang, and Roget's *thesaurus*."[42] To many, his style seemed difficult, opaque or odd. But it could be easily argued that Veblen's style constituted his ideas: a turn of phrase, a string of metaphors, or a salient expression offered new points of view. He would continually change course and switch discourse to outmaneuver his opponents. In his hands, words turned into powerful weapons that inspired generations of intellectuals.[43]

In his most influential books, *The Theory of the Leisure Class* (1899) and *The Engineers and the Price System* (1920), Veblen mobilized different nineteenth-century discourses for technology, with a particularly gendered twist. In *The Theory of the*

Leisure Class (1899), Veblen showed how the various cultural forms and institutions of the leisure class functioned economically, and ostensibly had merely a decorative value – as exhibited in the possession of dogs, the wearing of corsets, or the mounting of international expositions. In one of his more memorable passages, he argued, in effect ignoring the arguments of Tarbell and other women reformers, that women of the leisure class had primarily a decorative role, deriving economic value only by virtue of the men with whom they were associated. To the men of the leisure class, women's uselessness or idleness served as a token of men's leisure-class status. Woman "is man's chattel...she is useless and expensive, and is consequently valuable as evidence of pecuniary strength," for the man who "supported [her] in idleness."[44] Veblen opposed the notion that market relations determined value, because he believed that its true standard ought to be set by what is produced through socially useful labor; giving the economists' understanding an ironic twist, he showed how uselessness and idleness could turn into valuable and *useful* assets. For this twist, he became famous, of course. Nevertheless, he only reinforced that middle-class women were merely decorative because in the truest economic sense they failed to produce any goods. Theoretically, his definition did not consider engineers or managers as producers, but when confronted with a similar theoretical dilemma in the case of engineers, Veblen expanded the classical economists' definition of utility to include engineers under its label.[45] In his second book, *The Engineers and the Price System* (1920), Veblen classified engineers as valuable by designating them as producers. Having gone this far, he was challenged by the question: if engineers were producers, what indeed did they produce? Veblen argued that engineers were the actual producers of technical knowledge, or a thing he now called technology.

Veblen's strategy of portraying engineers as the sole bearers of technology is rather surprising. In his earlier work, he not only reserved a role for skilled workers but – more importantly in this stage of his intellectual development – during the first World War he wholeheartedly endorsed the goals of the Industrial Workers of the World (I.W.W.), who sought to claim technical knowledge for workers.[46] Inspired in part by Veblen's book, the I.W.W. leadership developed a theoretical position in November and December 1919 with regard to the use and ownership of "the joint stock of knowledge of past experience," and launched the idea of compiling a systematic "Industrial Encyclopedia" for workers. The I.W.W. called for all workers to join the effort in order to make a smooth and orderly transition from capitalism to socialism. The encyclopedia "would serve as a practical guide to the workers in fitting themselves to take over and run their industry."[47] Its initiators expected that the "joint stock of knowledge" would empower skilled and unskilled workers technically and would also prepare workers properly for the imminent takeover of the industrial system in the event of a revolution. For his part, Veblen also had intimate ties with the Wobblies and sympathy with the plight of women. Nevertheless, in the course of defending engineering knowledge, he began to obscure the control of skilled workers and to omit the women's tradition of Philadelphia's Emma Allison, the female textile workers of Lowell, or women's patent activities of Smith's sisters.

As a word, *technology* was the key to Veblen's argument presented in a series of articles for *The Dial* in 1919 when the air was filled with talk of revolution. He considered American engineers the only suitable candidates for leading a peaceful revolution that could unseat the vested interests of business monopolies and national unions because engineers belonged to a small, disinterested, and apolitical community: their only true interest was in the advancement of neutral technical knowledge and the working of the system, which he construed as a machine. As a General Staff of the industrial system, they could therefore best serve as the impartial, dispassionate caretakers of industry.

With his definition of technology, he forged a bridge between the nineteenth-century discourse of the industrial arts and the twentieth-century talk of technology. "Technology – the state of the industrial arts – which takes in effect in this mechanical industry," he wrote, "is in an eminent sense a joint stock of knowledge and experience held in common by the civilized peoples."[48] Here, Veblen referred to the old nineteenth-century meaning of technology as an inventory of industrial crafts that could be studied, but he also broached the idea that technology represented a disembodied object, devoid of any human agency. He now defined technology as an aggregate of knowledge and experience that could be held jointly, without exclusive rights to its ownership. In his formulation, technology was also an index to the level of civilization, as the anthropologists Morgan and Mason had argued.[49] Finally, Veblen pointed to mechanical industry as the locus of the "joint stock of knowledge and experience."

The designation of technology as an object resulted to a large extent from his frequent invocation of the machine as a metaphor – an image he exploited to the fullest, no doubt sensing the Dadaist vibrations also in the air at the time. Veblen, a historicist by training, often explained social phenomena in their institutional settings. Still, he made his unwitting contribution to the new economic and modernist language of machine efficiency that was devoid of social context. He spoke not only literally about mechanical engineering, but also metaphorically about the machine, to evoke both the industrial system and society at large. In his metaphorical language, he represented the industrial system as a self-generative and self-contained machine, where human beings were no longer needed. "The industrial system," he wrote, "is notably different from anything that has gone before. It is eminently a system, self-balanced and comprehensive; and it is a system of interlocking mechanical processes, rather than of skillful manipulation. It is mechani-

cal, rather than manual."[50] Independent as the machine might seem to be, Veblen
argued that production engineers would be needed at the helm and were the only
ones who could be entrusted with its supervision.

The notion of technology and the figure of the engineer entered Veblen's work
as an afterthought. It nevertheless proved to be a crucial one. His metaphorical
language enhanced a modernist mode but also suppressed important experiences
that did not fit the analogy. While he acknowledged the place of workers' skills, he
saw engineers as the chief bearers of technical knowledge. In the American language, Veblen was the first to use the term "technology" so frequently and lavishly.
He explicitly linked it with *engineers* and *productivity*. Throughout his life, he had
been concerned principally with monopoly capitalism and in particular with the
"corporation financier" as the embodiment of the non-producing classes, but his
main narrative strategy was to exploit engineers as a counterpoint to the corporation financier. To make this argument stick, however, he needed to extend the
economists' definition of what constituted productive labor, and also had to make
sure that engineers would be producers of a product. Earlier, he had not used the
word "technology," but the term was now liberally sprinkled throughout the pages
of *The Engineers and the Price System*, along with a host of machine metaphors.[51] As
the bearers of "the joint stock of knowledge of past experience," he insisted in this
1920 writing, engineers were producers of income. In Veblen's vocabulary, technology had become a product, and engineers were the producers of that product.
But Veblen had been listening to a minority position in the engineering profession. Ironically, different engineering communities – ranging from academic engineers, industrial researchers, and science-policy makers – still preferred the term
applied science over *technology*.[52] This would soon change.

TECHNOLOGY-AS-KEYWORD ON DISPLAY

Ten years after Veblen had argued that a government should be formed by technically competent leaders, his work sparked the short-lived Technocracy movement.
The discussion proved to be a watershed in the political alignment in the modern
discourse on technology. About 1930, "technology" became a buzzword incorporating anthropological notions of civilization, engineering professionalism, and
machine metaphors. It also became heavily invested with ideological weight when
the Technocracy movement captured the ongoing debate over the idea of technological unemployment in the thirties.[53] Technocracy helped popularize the notion
of technological unemployment, giving currency to the view that there was a problem with the current relationship between mechanization and work, as historian
Amy Bix argues. Supported by social scientists including William Ogburn, Stuart

Chase, and Elizabeth Baker, mainstream labor leaders like William Green of the American Federation of Labor and John L. Lewis of the United Mine Workers began to sound the alarm over what they called "technological unemployment": the displacement of labor by mechanization as a fundamental feature of industrial capitalism. Carefully avoiding any Luddite associations, labor leaders worried about mechanization's "human scrap-heap." They neither insisted on halting mechanization nor on suspending science research, but suggested that the burden should be carried evenly by the labor, business, and science communities. While several labor leaders and social scientists offered remedial measures like the reduction of hours and aid for displaced workers to soften the blow, many proponents of the Technocracy movement criticized or opposed industrial capitalism altogether. The Technocracy movement offered a Veblenesque solution to the perceived problem: the restructuring of the price system and the government by engineers and scientists. According to its powerful opponents, however, the claim of the Technocracy movement to engineering knowledge was a damaging and onerous one that demanded forceful rhetorical answers.[54]

The established engineering community disavowed the Technocracy movement with exceptional ferocity despite the many personal and intellectual links between the Technocracy movement and engineering – especially the newer, corporate, and laboratory-oriented branches like electrical engineering during the Depression. Business leaders, research scientists, and academic engineers including Karl Compton, Robert Millikan, Michael Pupin, Charles Kettering, Frank Jewett, and Arthur Little quickly closed ranks on the issue. The corporations aggressively sponsored the 1933 Chicago and 1939 New York World's Fairs as part of an elaborate public relations campaign to divert attention away from the discussion about the issue of technological unemployment and the unemployment among engineers. The 1933 and 1939 displays shifted the focus from the early interest on producers and their products to the wonders of consumption instead.[55] Likewise, establishment engineers and scientists balked at the suggestion that they should be blamed for the human misery. In this controversy, the rhetorical use of the term "technology" proved to be essential. It had been offensive enough to suggest that scientists and engineers were responsible for the human misery of the Depression, but the mere suggestion that Technocracy's leadership laid claim to the mantle of engineering knowledge to advance its radical agenda was even more unsettling. Business leaders, establishment scientists, and engineers swiftly mobilized by denying the charges, ridiculing the movement, and insisting that scientific advance, economic success, and the progress of civilization were indisputably linked.

The mobilization of the notion of technology against Technocracy served a rhetorical purpose. Responding to the claims of the Technocracy movement, physicist, President of MIT, and public spokesman for the science and engineering community Karl T. Compton (1887-1954) wrote in *Technology's Answer to Technocracy* that neither the movement, nor its analysis of an economic crisis, nor its concern for workers displaced by labor-saving machinery amounted to anything new.[56] As the rank-and-file members of the profession experienced a decline of wages by a third as well as bouts of unemployment, the engineering establishment reacted with ferocity to the association of their profession with technological unemployment and to the 'misuse' of engineering credentials by many of Technocracy's proponents, whom they labelled "pseudo-engineers" and "quacks." In 1933, Arthur Sheridan, a former president of the New York State Society of Professional Engineers, blamed political scientists for trying "to place Technocracy upon the doorstep of engineering" and "seeking to discredit engineering as a social factor in civilization through condemnation of Technocracy."[57] Arthur D. Little (1863-1935), the industrial chemist, spokesman of engineering professionalism, and founder of the oldest and best-established firm in research and development contracting, contested even more sharply the claims of the Technocracy movement and the use of the word *Technocracy*. "In happy contrast to the gloomy futilities of Technocracy stand the solid achievements of that very different thing, technology. In a little more than 100 years technology has increased, immeasurably, the wealth of the world...Technocracy is destructive; technology is creative. Let us not confuse them."[58] He thus linked Technocracy to the pessimism of the Depression, and technology with progress.

The public answers of Compton, Sheridan, and Little disavowed the revolutionary role Veblen had suggested for engineers, denied any links between engineers and unemployment issues, and above all reclaimed technology for engineering experts. As Veblen had ironically anticipated, Compton and Little resolutely aligned technology with a conservative agenda, cleansing it of any anti-capitalist contaminations and pro-labor associations, and casting engineers and scientists as producers of wealth. If, as Bix has argued, in the short run, the Technocracy movement popularized the discussion over technological unemployment, in the long run, it harmed the case of those willing to argue that technological unemployment posed a serious problem. Engineers like Sheridan, Little, and others did much to define and protect the boundaries of engineering. Claiming semantic ownership of the term served this effort.[59] The discourse on technological unemployment and the fate of the Technocracy movement proved to be crucial factors in determining who could claim the true parentage of technology. The question of who owned technology, or who could claim its progeny, fundamentally redirected the discussion.

In the 1930's, social scientists like the University of Chicago's William F. Ogburn reworked Veblen's oeuvre and the anthropological tradition of Morgan, granting scientists and engineers an active role as agents of history over and against politicians and statesmen. They elaborated on Mason's genealogy of invention even further in merging the two discourses of academic engineering and cultural anthropology. In Abbott Payton Usher's words, scientists, engineers, and inventors were twentieth-century incarnations of Carlyle's heroes: small, anonymous, but essential.[60] These early sociologists of invention rejected biological racism and firmly agreed with cultural anthropologists like Franz Boas and Alfred Kroeber that inventions could occur simultaneously in different cultural settings and were seldom creations of an inventive genius. But in their cross-cultural comparisons between primitive and civilized cultures, they also reiterated that inventions – understood as mechanical and patented – explained the difference.[61] When faced with an explanatory gap between the premise that all cultures shared the same human nature and the notion that in a short period of time the Western world had generated many patented inventions, theorists of invention including Ogburn, Usher, S. Colum Gillifan, and the popularizer Waldemar Kaemffert – the science editor of *Scientific American, Popular Science Monthly*, and *The New York Times* – allotted a crucial place to inventors with extra intelligence: the great men in history. More explicitly than Veblen, these sociologists turned engineers into male heroes using images from popular fiction. Like Mason, they recoiled from taking their monogenetic theory to its logical conclusion. They included neither other cultures nor women in their theory to explain the difference between Western and other cultures, but fixed scientists and engineers as male agents of history, who produced what they now called technology.

The rise of the National Research Council (NRC) and the Departments of Commerce after the first World War surpassed the Smithsonian in classifying and in laying claims to the objects of the twentieth century. Up to the first World War the Smithsonian had been instrumental in fostering the material world that still could include a range of inventions from clay pots, and corsets to cars, but when the NRC joined hands with corporate sponsors in organizing the 1930s World's Fairs it legitimized a new language of technology closely associated with industry, the military, and the professionalizing communities of science and engineering. Established by Woodrow Wilson, the NRC would become the primary agency for promoting the cooperation of science, industry, and the military. Big science and big engineering became ever more closely tied to the *military-corporate* complex. If the Smithsonian-organized fairs still included a hodgepodge of artifacts from clay pots, and bonnets to reapers, the NRC's staged affairs resolutely resembled clean corporate machines. More importantly, the rise to prominence of the NRC in staging the World's Fairs shaped the emergence of a new paradigm that put scientists, engineers, and corporations center stage as the producers of technological artifacts and cast women, workers, and African Americans as consumers.[62] In the course of a century, technology had been turned into a product, engineers into

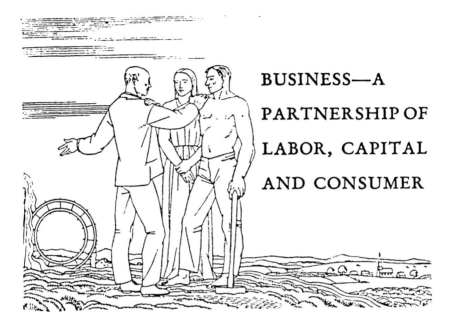

Figure 8.   Corporate allegory of a manly partnership between Labor and Capital witnessed by
a female Consumer. Reproduced from autobiography of engineer and captain of industry Paul
Weeks Litchfield, *Autumn Leaves: Reflections of an Industrial Lieutenant* (1945).

producers, and women and workers into consumers who were mere onlookers of
the technical enterprise. As time went by, technology would come to mean the history of corporate engineering. [Figure 8]

2

From Elite Profession to Mass Occupation

Engineers emerged as the shock troops of industrial capitalism. Nevertheless there was something curious about the engineer's success to command male cultural authority. While intellectuals, artists, and social scientists endowed them with great cultural meaning and importance, many engineers felt misunderstood, disrespected, and undervalued. They suffered from existential anxieties what it meant to be an engineer and where the boundaries of its knowledge domain lay – anxieties that came to the forefront at three historical junctures.

From the World's Columbian Exposition in 1893 through the aftermath of the first World War, many advocates for the engineering profession argued over the definition of engineer in countless articles that appeared in the technical press. These spokesmen for the occupation included engineering educators and reformers, editors of technical journals, and leaders of occupational organizations. Coming on the heels of bitter industrial struggles, their articles expressed intense status anxieties of patrician engineers who faced an aggressive industrial development and who felt threatened by a sense of class war and a menacing procession of immigrant cultures. In this setting, contests over the criteria for membership of the professional organizations between the various factions within engineering did not merely define the term engineer. They helped claim a special knowledge for engineers and legitimized their cultural authority coded in terms of a revitalized manliness. The leading American bridge engineer John A.L. Waddell (1854-1938), born of Irish-American parentage and closely associated with America's imperial projects at home and abroad, had built the larger part of his career on overseas work in Mexico, Cuba, New Zealand, Russia, and Japan. Waddell liked to lecture students and his peers on the many tactics of upgrading the status of engineers and expressed his concerns about the proper class, gender, and ethnic boundaries of the engineering profession before a circle of academic engineers in 1903 thus: "We have the man who fires the boiler and pulls the throttle dubbed a locomotive or stationary engineer; we have the woman who fires the stove and cooks the dinner dubbed the domestic engineer; and it will not be long before the barefooted African, who pounds the mud into the brick models, will be calling himself a ceramic engineer."¹ Through his figure of speech, Waddell belittled the skills of mechanics, women home economists, and bricklayers as a rhetorical strategy that sought to upgrade the profession. He incorporated a scientific racism that became more coherent and articulated in the decades after the World's Columbian Exposition in 1893: it was a racism that elected James Watt rather than Mason's African women as early technologists. Establishment engineers – among whom academic engineers were most vocal – insisted that the distinction and the term engineer be respected, as if to counter the occupational ambiguity that always seemed to be lurking in the background.

American engineering transformed from an elite to a mass occupation, grew the fastest of all, and differentiated at a great pace from the 1890s onwards. It was also a deeply divided and segmented profession whose practitioners could be found working anywhere from boardrooms to drafting departments, mechanics workshops, and chemical labs as executives, managers, designers, draftsmen, detailers, checkers, tracers, and testing technicians: by 1935 there were 2,518 different engineering job titles. Lacking the classic gatekeeping mechanism of a central agency, the state, or professional organization, American engineering neither became a closed profession associated with science as in France where the state groomed a small elite for leadership positions; nor did it fashion itself after the British engineering culture of small family firms, craft traditions, working-class associations, and kinships. Instead, American engineering would evolve into something between the French and British models: a mass middle-class occupation with a hybrid form of professionalism and an almost knee-jerk aversion against classical blue-collar unionism.[2] When the gates were opened to newcomers from lower-class and different ethnic backgrounds, maintaining the middle-class position proved to be tenuous, however. In an immigrant society where upward mobility marked a bone of contention, the rhetorical positions on manliness and race often masked the tensions of class.

Engineering advocates were engaged in a balancing act of maintaining the class status of the profession when it transformed from a small elite profession to a mass occupation. Nineteenth-century engineers, many of whom had been employed by the railroad corporations, formed a class of middle managers who invented, innovated, and arbitrated in the emerging federal and corporate bureaucracies. When the state and the corporations expanded dramatically in the early twentieth century, new generations of engineers – many now from ethnic backgrounds – started to fill positions in the lower rungs of the middle class.[3] As a group, engineers found themselves not merely caught in the middle, they also acted as an active and self-conscious constituent as a middle class in formation.[4] Establishment engineers like Waddell tried to advance their lofty professional ideas within the pristine walls of engineering schools, the pages of technical journals, or the halls of professional organizations. Educational standards did not manage to lay a foundation for a monopoly or serve as a benchmark of competence. Subsequent movements of engineering reformers failed to introduce a professionalism along the lines of their medical brothers. Closely associated with industry, establishment engineers rejected uniform education and credentialing rules. The efforts of Progressive engineering reformers like Morris Cooke to sever the ties with the American business community failed during the pivotal period between the 1910s and 1930s when the discontent came to the fore. Nor were academic engineers successful in restricting access to their ranks through education.[5]

The insistence on a proper definition of the field stemmed in part from the ambiguous and loose nature of engineering, the inability to find clear gatekeeping mechanisms. At a time when Waddell worried about the proper race and gender boundaries of the profession, the fastest growing occupation transformed into a mass career for many. In the balancing act, gender and race were more rigidly maintained in a society where upward mobility marked a bone of contention. Racial and gender lines were therefore more carefully drawn. Often the rhetorical positions on manliness and race masked the tensions of class that were part and parcel of the most greatly expanding occupation of all. Waddell's rhetorical position fell squarely with the leading engineers's search for professionalization.

In the negotiations over boundaries of class, three major cultural conflicts shaped engineering that sought to renew its middle class character through a language of manliness and whiteness. The first cultural conflict erupted over the question of where the true path lay toward an engineering career and technical expertise and knowledge: through the doors of the rough workshop or the genteel schoolroom. It pitted propriety engineers and academic engineers against each other in a struggle over renewing male authority based on class relations of the workplace or on science. This contest, at its height during the 1890s, has entered the secondary literature as the tensions between the shopfloor versus the school culture. In the decades that followed, establishment engineers and the rank-and-file engineers struggled over the direction of professionalism and engineering unionism. In this struggle, the danger of being declassed and demasculinized threatened the status of the engineer as a middle-class man. Finally, as the government-military complex gradually emerged as a result of the first World War, academic scientists and industrial researchers like Karl Compton, Robert Millikan, Arthur Sheridan, and Arthur D. Little argued about the meanings of applied science and technology. The rhetorical contest centered on the corporate sponsorships, federal research and development resources, and the engineer's proper cultural authority and status as a corporate man. The contest culminated in the struggles over the meaning of technology during the 1930s that recaptured and reworked earlier episodes and came to mean how we understand technology today: a white, middle class, and male enterprise that sought to claim an exclusive expertise for engineers over other forms of technical knowledge while at the same time making universal claims for it. Although each cultural conflict has received attention, they have been considered neither in terms of class, gender, and ethnicity nor in relation to each other. When immigrant and lower-class sons began to enter through the system of American engineering education, the boundary work of class became an ongoing process of redefinition, while at the same time gender and race boundaries were upheld. The sons of the lower middle class and immigrants joined, but, with few exceptions women and African Americans were kept on separate educational and employment tracks. The boundary work of class was reworked, rejuvenated, and remade through these three historical episodes.

'Shopfloor Culture' and the Workplace as Moral Gymnasium

In the middle of the nineteenth century when America laid a grid over the newly conquered western territories with canals, turnpikes, and railroad tracks, and urban centers sprang up along the way, engineering offered aspiring men social status and, later, the promise of upward mobility. It promised a more secure income and opportunities for advancement to a young man who had decided the time had come to be a serious breadwinner. Alfred West Gilbert (1816-1900), the son of a tenant farmer, explained how he settled on engineering when he fell in love with his future wife and saw the rapidly expanding public works projects in the mid-Atlantic region.[6] After his father offered to pay for his education, Gilbert became a city engineer, surveyor, and lawyer in Cincinnatti, a city then emerging on the banks of the Ohio river – the center of steamboat building and repair, home of many steamboat-related industries of general machine work and machine-tool manufacture. His career was closely associated with the building of the city's water infrastructure. In a similar fashion and of the same generation, James Worrall (1812-1885), the son of an Irish-born bookseller in the manufacturing city of Philadelphia, aspired to the life of genteel culture but abhorred the idea that intellectual ideas could be sold as books just like any other commercial goods. He wondered after meeting his future wife how he could escape his father's business and find a job, and decided on engineering in the 1840s when "everywhere around me I saw constructors, builders of churches, of wharves, of canals, of fortifications."[7] Some men went into engineering to affirm or replicate their comfortable backgrounds. Others – like the sons of impoverished Southern plantation owners who became the shock troops of the New South after the Civil War – did so in an effort to recapture their social status after family fortunes had been lost. In America, engineering was an occupation of the middle class. Unlike France, in the U.S. shared expectations, aspirations, and goals – not family backgrounds – came to shape the social cohesion of engineering since it was linked to industrial capitalism rather than state service. In America, engineering did not reproduce old elites or middle classes; it groomed young men into a sense of class through distinct rituals, narratives, and self-representations when it expanded dramatically.

During the first half of the nineteenth century, it was quite common for engineers to be trained exclusively on the job. In the absence of engineering schools it had been standard practice on Erie canal's building sites to recruiting engineers by promoting capable chiefs of survey crews. This kind of informal recruitment and apprenticeship carried over from America's internal improvement projects to other sectors including the building of the railroads. The building of the first major American railroad, the Baltimore & Ohio, served as a laboratory for training many civil engineers, and a generation later the construction of many railroad bridges spanning the major arteries gave many engineers the necessary experience to become bridge specialists, later used in countries as far away as Uganda. In the gold and silver fields of California and Nevada, engineers learned new techniques in deep gravel and hard-rock mining. Two generations of mining and civil engineers, following the tracks of the federal government's topographical engineers since 1824, extended the American sphere of influence, ventured into the American West in the 1840s only to move into Mexico, Cuba, Panama, the Philippines, Australia, and South Africa from the 1870s through 1890s; they sampled ore, staked out claims, supervised trial drilling, furnished drawings, calculated estimates, and acted as promoters of mines on behalf of investors with interested stockholders or as managers in hiring immigrant or foreign labor.[8] On-the-job training on the construction sites of canals, railroads, and urban centers became a common method of producing civil and mining engineers. Here aspiring young men worked in sex-segregated workplaces and shantytowns alongside common laborers and skilled workers from England, Ireland, Scotland, Germany, Italy, and Mexico from whom they expected to learn all the aspects of their trade in order to move up and out of the lower ranks as soon as possible. Starting as laborers responsible for clearing land alongside the Irish and slave African-American workers, on the railroad and canal building sites, they worked with the expectation that they would move up the ladder from chainman to rodman, to transitman, to surveyor, and eventually to assistant engineer. The labor camps along the tracing lines of projected canals and train tracks were largely societies of men, where hard living, hard working, and hard drinking were cherished values, reminding many an aspiring engineer of the kind of proletarian manhood they were determined to avoid at all costs. Always terrains of labor conflicts over wages, working conditions, and control, emerging unscathed from the rough and tumble culture of the labor camps became rites and sites of passage into manhood shaped by a contest over class relations. "Cannallers," writes their historian Peter Way, "participated in sport, fighting, boozing and various contests of strength, in the process developing a proletarian sense of virility and physical prowess."[9] Rough amusement of drinking, laughter, and boisterous fighting expressed the deepest solidarities and resentments of lower-class men, who came to embrace this rough code of manhood as a way of deriding and resisting respectable and moralistic manliness of the middle class into which the engineers entered.

In this context, practical training was not only an important channel for the formation and social reproduction of middle-class identity, it also represented a formalized ritual of male socialization of middle-class men.[10] Born into a family of lawyers, Robert Ridgway (1862-1938), who would become an important urban engineer of teeming New York later in the century, had received only a minimum of academic instruction because the panic of 1873 forced his lawyer father to move his children back to the family farm. Instead of going to college, Ridgway went off west to the field-school of engineering – the expansionist projects of the large railroad construction in the West. The young Ridgway rode the wave of the railroad building boom in the period between the economic busts of 1873 and 1893, entering upon his engineering career at the age of 20 as a rodman on a surveying expedition in Montana during the summer of 1882, only to continue with the Northern Pacific Railroad in Wisconsin. He received his first engineering training on the building sites and in the labor camps in the newly acquired Western Territories in Montana where he hoped to find his manhood. After his initiation into manhood in the West, he returned East because "the frontier today is in the cities – not on the prairies." He built his career by participating in the new engineering infrastructure of streets, sewers, water supply, tunnels, electrical lines, and subway systems that were laid out for the booming urban center of New York – the port of entry for millions of immigrants at the turn of the century. He obtained his first promotion to the position of leveler three years later, when he joined the building of New York State's aqueduct system at Croton. There, he had an explosive conflict with the Italian masons over the lining technique of the tunnelling. The conflict between engineering knowledge and the skilled Italian masons was a classic contest over who possessed the best technical knowledge and controlled the workplace. Ridgway and his engineering corps won. Ridgway worked ten more years before moving up from senior assistant engineer in New York City's Rapid Transit System building projects to division engineer on the New York East River Tunnel construction works in 1900.

It would take more than twenty years of informal recruitment procedures, on-the-job training, and careful negotiation with workers and contractors over expertise, command, and control on the building sites from Montana to New York before Ridgway could call himself an engineer.[11] To help negotiate the tension be-

Figure 9.   Civil engineer Robert Ridgway posing in front of the field office near Old Croton Dam in 1887 and seated top step on the left amidst young engineering trainees. Note the skilled Italian mason at the extreme right whose knowledge of tunneling technique Ridgway and his engineering crew contested. Reproduced from Robert Ridgway, "My Apprentice Days," *Civil Engineering* (1938), Courtesy of Delft University of Technology, Delft, The Netherlands.

tween capital and labor, between field engineers and skilled workers, between white collar and blue collar, Ridgway's generation dressed for their outdoor activities on the building site while keeping a safe distance from their working-class subordinates through carefully chosen headgear and shoes. [Figure 9] Well into the twentieth century, this kind of on-the-job training made up an important segment of the engineer's overall schooling. In the textile industry, practical training also known as the "tour of the mill" was still the centerpiece of mechanical engineering training and rounded off a formal education until the 1890s. In the nineteenth and early twentieth centuries, proprietors' sons were no longer expected to master a trade, but because of the increasing complexity of the manufacturing process they still trained in the plant to acquire a working knowledge of all the facets of the firm, including the materials used, processing, production, sales, and repair.[12]

The belief in and practice of shopfloor knowledge became canonized in the notion of the importance, if not the necessity, of hands-on and field training reproducing for a new age a patriarchal authority that was based on class relations rather than on science. The ideological base for this shopfloor culture came first of all from the machineshops in Pennsylvania. In Philadelphia and other eastern industrial centers, advocates and managers of the shop culture formed a class-conscious elite with a network of family connections resembling British recruitment patterns. Many sons of this culture of family-owned business found themselves in a

Figure 10.   Bethlehem Steel's Machine Shop no. 2, site of Frederick Taylor's metal-cutting experiments around 1900. Mechanical engineers' moral gymnasium framed as a production floor without workers. Courtesy of Hagley Museum and Library, Wilmington, DE.

period of transition when their opportunities for proprietorship was shaped by ill-defined, middle-level positions in large organizations. Working at the Baldwin Locomotive engineering shop, Raffe Emerson, the son of Taylor disciple Harrington Emerson thought the experience would make a man of him physically, something he believed mere strenuous exercise would not do. As he wrote to his father in 1904, "A few months of this kind of work will...be the best kind of a brain reast [sic] and body builder for me. Sailing a boat, or hunting might be somewhat more healthful, but not nearly so profitable as experiences."[13] Young men of Emerson's circle cherished Emerson's brand of manliness that valued bodily prowess, individualistic assertiveness, and valuable experience as measures of manliness. The machinists Emerson worked with, by contrast, defined their manliness in terms of skill and solidarity. These nineteenth-century male working codes not only served as an initiation rite for the profession, but for middle-class manliness in the 1890s. For sons raised on a work ethic, the accumulated riches of successful fathers posed a threat to proper character development. The aura of independence, the esprit de corps, and the sense of service to society embodied in the ethos of professionalization offered an acceptable way out of the dilemma.[14]

For men of Frederick W. Taylor's generation, the factory and the field constituted a moral gymnasium for the exercise of character development. [Figure 10] Taylor (1856-1915) was not the only one of his generation and class who, after failing health, would abandon his classical education and turn to the field of engineering. After Reconstruction, many of his contemporaries treated the field or the shop as a site where they could reconfirm their quintessential manliness defined by class.[15] In arguing over what constituted the true mechanical engineer, members of the ASME discussed where the honor of engineering lay or what distinguished them from stationary, locomotive, or marine engineers – the kind of skilled workers that Waddell also contested. In 1895, one member "disavowed meaning any disrespect to overalls or greasy hands, and said he had worn overalls and had his hand and face as black as anybody else ever had, and was not in the least ashamed of it."[16] Nevertheless, such provisional dress was carefully coded to flag class distinctions in the workplace.

The rites of passages of grease, overalls, and blackface on production floors, building sites, and in mine shafts were closely linked to the network of kinship connections. For most of the nineteenth century male socialization resulted from paternal relations between fathers and sons, which developed in tandem with the fraternal relationships between fathers and their peers. These family networks and informal relations took their most institutionalized form in the more elite engineering societies and provided the sons with a platform for becoming engineers and advancing in their career. The well-traveled mining engineer, author and editor Thomas Rickard (b.1864), who boasted an international career in the mines of Colorado, France, Australia, and New Zealand, relied on his extensive experience when as late as 1906 he still advised that a good start "such as a father able to pay for the necessary education, a kind uncle to give the graduate a job, and friends glad to give a push when most needed" was vital for success.[17]

This pattern of socialization and acquisition of technical knowledge occurred not only in small proprietary firms but continued in the emerging corporations that were based on the patriarchal culture of family firms. Even when the chemical firm DuPont Company went through a large organizational transformation from a Delaware family-run firm to a major vertically integrated corporation, expanding from 1,500 to 60,000 employees from 1902 to 1915, the old social networks and functions still guided some aspects of the managerial routines. As mining engineer Thomas Rickard had advised earlier, Harry Pennington turned to his social peers on behalf of his nephew to request another position for him because of the arduous nature of the work he faced on his first job as a day laborer at the DuPont Company in 1915. In an apologetic response, DuPont's General Manager Hamilton Barksdale arranged a less strenuous job for Pennington's nephew, but he pointed out, "he is, of course, at the present time filling a minor position, as his lack of experience necessitates, but I should say that his future advancement depends only upon his making good the impression he has already created."[18] Barksdale found the assumption that Pennington could interfere quite acceptable, but also cited his nephew's lack of skill to underscore the rules of the system. The system combined paternal protection with the principles of meritocracy or "making good." Paternal protection meant that seniors would recognize, encourage, and foster the aspirations of a young man who started at the bottom of the corporate ladder as a day laborer. To a certain degree, plant experiences were vestiges of an apprenticeship system that had sought to integrate young males into the ranks of skilled workers, and had symbolized their passage into their adult roles as husbands and fathers. These rites of passages of the workplace revitalized a middle-class manliness for industrial capitalism when old patterns of male socialization had been broken. Such passages denied, but also resolved differences of class.

On the shop floors of the steel industry in Philadelphia, in the mine shafts of Colorado, along the railroad tracks in Montana, in the drafting departments in New York and at the labs in Dupont, engineers encountered a host of other groups of workers like powerful iron puddlers, independent miners, unruly canallers, low-level clerks, and ill-paid chemical analysts. They competed with these male workers for authority and control, and sought to differentiate themselves from them as well, but – and this was the crux of the matter – the engineers also depended on these workers as they needed to learn the tricks of the trade from them.

As engineers became increasingly channeled into white-collar jobs as managers, "rubbing shoulders" with the men on production floors and building sites assumed a particularly symbolic meaning as markers of manliness. Engineers in managerial positions for the most part sought to control the process of production and the performance of workers in the plant, but their association with working-class manliness also "rubbed off," as it were, onto them. Their authority over other men depended on an ability to speak their language *with* them without becoming *of* them.[19] The true engineer neither simply employed the chemical formulae once memorized at school nor learned the technical aspects of production, but perhaps more importantly, also sought to command a cultural fluency of the workplace – the men who could talk shop with the workers, while simultaneously supervising and keeping a safe distance or, to put it in the words of civil engineer Otto Garman, the facility to "adjust your vocabulary to the shop, the camp or the trench [without forgetting] that there is a different vocabulary and suitable subjects to be used in the drawing room."[20] In order to gain authority over the Irish, Welsh, Scottish, Canadian, and Italian workers they supervised, management engineers needed to be able to speak the language of the workplace. In the case of skilled workers, this learning process rested on the willingness and cooperation of workers to share their knowledge. Unlike the fields of medicine and law, socialization in engineering occurred through the association with and the simultaneous rejection of the manhood of the class of lower professionals, skilled workers, and common laborers with whom engineers worked in the workplace. Abstaining from alcohol and proper dress became important markers of class difference.

When engineering changed from an elite profession to a mass occupation and when management practices sought to reduce skilled workers to mere physical laborers by extracting skills and by removing work decisions from production floors and building sites and placing them in the hands of personnel officers and middle managers, middle-class men became even more heavily invested in the ideal of bodily prowess and hands-on experience. The romance of the shop floor, its cultivation of stamina, and bodily prowess gained ideological expression at the very time that management practices started to separate manual from mental labor and engineers became more removed from the tough talk on the production floors and building sites.[21] This ideal of manliness located manly characteristics in the body and in individual achievement. In contrast to the late eighteenth century's emphasis on refinement and status in the community or the body politic, the emphasis on the male body found its most forceful expression in upper-class men's celebration of sports like basketball, volleyball and rugby.[22] This emphasis had a particular meaning for management engineers. While men like Taylor, Raffe Emerson, Ridgway, and John Fritz celebrated their feelings of fraternity with skilled workers, they also participated actively in reducing and controlling workers' skills. In recalling their work days, engineers portrayed themselves as active builders when they merely had the supervision over others carrying out the actual construction work. Thus, a switch occurred in both language and image. Management engineers and engineering educators appropriated not only the mental labor of skilled workers, but also metaphorically appropriated their bodies. Rather than speaking of their supervisory functions, they seized the language of building and design as the tangible result of their own engineering creativity.

The association of engineering's identity with rough-and-tumble and hands-on experience of production floors and building sites was both real and idealized. Business engineers came to articulate their shopfloor identity when they became overwhelmed by the scientific claims of academic engineers and by the unions' defense of workers' technical knowledge against management encroachments. The affection that engineering managers cherished for their workers had distinct limits for they showed unmistakable hostility to the desire of skilled workers like British mechanics, Welsh miners, and Italian masons for control over their work, and their aspirations to have their own unions. Only rarely did successful engineers show support for workers' unions, or, for that matter, for engineers' unions. When lower ranking engineers articulated their own desire for an organization that would address their pay scales and working conditions, telephone engineer Garri-

son Babcock (b. 1879) insisted that the new organization of engineers "must overcome the impression that because we are an association of engineers we are a
union."[23] Engineers had a knee-jerk aversion against blue-collar unionism. If a
skilled industrial workers' organization based its strength on economic and social
solidarity among workers, the social cohesion of engineering rested on the promise
of individual social upward mobility and the simultaneous acknowledgment of
such aspirations by men in the higher echelons.

'School Culture' and the Domestication of Outsiders

Those who celebrated the production-floor manliness of grease also responded to
other contenders of their technical authority. After the Civil War, proponents of
formal schooling began to advocate a contrasting vision of engineering skills. Engineering schools openly challenged the experience-based education so firmly entrenched in the industrial craft traditions in the machine shops, mine shafts, and
on the building sites of the canal, turnpike, and railroad projects throughout the
U.S. and beyond. Catering to the new demand for young recruits to staff the industrial expansion, academic engineers sought credentials to monitor admission
into the engineering ranks. They claimed to offer their students "methodical and
scientific application to everyday actions," instead of the unregulated and despised
rule-of-thumb method.[24] Within the engineering establishment, the conflict between the two models of education – one based on formal education, the other on
on-the-job-training – came to a head during the 1880s and 1890s. Despite the
growing numbers of school-educated engineers over the following half-century,
however, as late as the 1940s about 45 percent of male engineers and chemists who
worked in or had applied for engineering positions had yet to complete their college educations.[25] Only after the second World War would engineering schools
win this struggle over credentialling and acquire the kind of male privilege the old
patriarchal culture of the family firms had once possessed.

On the eve of the Columbian Exposition, academic engineers began to rally
forces in establishing in 1891 the Society for the Promotion of Engineering Education. It was before this organization that bridge engineer and the profession's
prominent promoter Waddell entertained his brethren by ridiculing railroad driv-
ers, women home economists, and "the barefooted African, who pounds the mud
into the brick models," but called themselves engineers. His sneer was less a play
on words as it appeared at first, than a rhetorical strategy that sought to fix clear
class, gender, and race boundaries at a time when American education was expanding dramatically to extend its promises to lower class youths and a few

women and African-American graduates who scaled the engineering bastion. Waddell drew on several vocabularies and reworked it into a new discourse.

In the decade following the Civil War, diversity and openness characterized American engineering education. It nevertheless came to be bound by gender and race. Hailed as the landmark legislation aimed at pushing higher education to unprecedented levels, the Morrill Land Grant Act of 1862 helped establish several schools of engineering at land-grant state universities, colleges, polytechnic institutes, and private universities throughout the land. During the Civil War, the Northern Congress had legislated the Act for the express purpose of educating the children of farmers and industrial workers. Its drafters had not stipulated with precision what they meant by "agricultural and mechanic arts," however, and in the early days, women, workers, and farmers attended courses given at institutions like MIT. Industrialists who had sponsored the Act not only had been the first in supporting education of the "industrial classes", they had also been at the forefront in sponsoring the instruction of women's in a temporary alliance with women's rights activists. As sponsors of the "useful arts," they welcomed the enrollment of women, viewing them as a potential disciplined workforce.

This broad commitment continued in the early days of the Morrill Act, but this would change in the course of the century. Under the aegis of members of Congress from the North, state officials, local business men, and engineering educators "the agricultural and mechanic arts" often came to mean industrial rather than agricultural education, technical rather than artisanal training, and school-based engineering rather than a British-style apprenticeship. More often than not the allocations came to respond to the need for engineers in the development of industrial centers in the East or in the exploitation of the rich mineral resources and the expansion for mining and railroad lobbies in the Western Territories.

Over time, educators and local business men began to upgrade the various industrial education programs and instituted policies which kept at bay those groups that had been the bill's main mission initially. The push of upgrading the field through the infusion of professional ideals resulted in the masculinization of higher education of engineering in the U.S. that historian Margaret Rossiter has demonstrated for the sciences as well. Leading engineering educators affiliated with the Massachusetts, the California, and the Stevens Institutes of Technology tried to formalize engineering knowledge by linking it with scientific tradition, or what they came to call applied science. Breaking with the traditions of vocational training, these advocates managed to seize all the attributes of scientific rhetoric.

The term applied science seized the older terms useful arts and mechanic arts and imbued it with the cachet of science. After the depression of 1873, MIT's Francis Walker and his crew of young professors thoroughly transformed the institution from the original design of founder William Roger and his abolitionist and unionist circle. They closed down the school of mechanic arts and deemphasized popular lectures and the society of the arts that catered to the local community. In this effort, the terms applied science and technology marked rhetorical positions that allowed them to cleanse and sanitize the profession from the sweat, dirt, and calloused associated with labor. Jacob Bigelow's title of his chair at Harvard, "application of science to the useful arts", expressed that middling position between the skilled workers of the mechanics institutes and Harvard's gentlemen of science. *Technology*, as MIT students sang in their college songs at the end of the century, was founded on science and art, not on the skill or experience associated with workmen. And with an eye to the gentlemen then residing across the Charles River, the reformers of technical education at MIT eventually appropriated the idiom of science to drive a wedge between Harvard's genteel education for the "gentlemen of science" and the more vocational and shopfloor-oriented instruction advocated by the mechanics' institutes. In the process of professionalization, physicists had dropped topics like hydraulics and mechanics, while engineering advocates appropriated them. Both sanitized their fields from any working-class associations.[26]

Even if the rhetoric of science began to enter the walls of some leading engineering schools, the funds such institutions and the hodgepodge of other engineering schools spent on scientific research did not reflect this ideal at all. Shifting coalitions between practical engineers, science-oriented academics, and curriculum reformers battled over the direction of education all over the country in the period between 1880 until the second World War. Pecuniary reasons also protracted the ideal from becoming reality. Scientific research at engineering schools would become an option only after 1940 when the federal government earmarked large sums for research as a result of major wartime allocations.[27] The rhetorical position in mid-century anticipated rather than followed social reality, yet it served a clear purpose. From Boston to Hoboken, leading academic engineers began severing technical education's intimate ties to local communities and shedding working-class roots and its alliance with women's rights' advocates. If in the early years of building an institution most engineering programs admitted a few women as special students to increase enrollment numbers, once the educators felt on firmer ground, they started to spurn them as part of what Margaret Rossiter has called the masculinization of the professions. Significantly too, the co-educational land-grant institutions were the first to sponsor home economics as separate career tracks for women which diverted many rural women, interested in technical fields.[28]

The chief architect of the ideology of "school culture" and shaper of the professional boundaries of mechanical engineering was engineering educator Robert H. Thurston (1839-1903). Thurston was descended from the early settlers of Rhode Is-

Figure 11.   Cartoon of civil engineering students at the Land-grant institution Purdue University presenting themselves as individual bridge builders working in the field rather than as participants in large labor-intensive projects. Reproduced from engineering class book *Debris* of 1893.

land as the eldest son of a prominent manufacturer of steam engines, but had no knack for business to follow in his father's footsteps. Instead, he pursued an academic career as a professor at Stevens Institute of Technology and at the land-grant institution Cornell University, where he inaugurated the first mechanical laboratory and shop courses in the country, and became the first president of the ASME in 1880, and a member of the AAAS. He developed his argument regarding the academic aspect of engineering knowledge ("the application of scientific theories to the useful arts") in a debate with a Johns Hopkins physics professor, Henry Rowland, who espoused the notion that science rather than engineering stood at the pinnacle of true knowledge. Carving out a space between advocates of pure science and those extolling the virtues of practical engineering as a moral gymnasium, academic engineers like Thurston never grew tired of lobbying for formal technical education as the true path towards engineering knowledge.[29] In the hope of achieving their aims, academic engineers increasingly relied on tactics of professionalization and adjusted their educational goal forging ties with national industrial corporations and visiting local factories with their students.

Nevertheless, graduates from engineering schools found themselves unprepared for the resistance from seasoned field engineers, powerful employers, autocratic foremen, and skilled workers alike, who had little use for academic attitudes and continued to put their faith in knowledge learned through their craft's tradition, their own experience, or existing plant routines, and who accomplished innovation through a method of trial and error rather than through the scientific procedures advocated by the academicians.[Figure 11] The experienced field engineer Ainsworth remembered the four engineering students and an assistant professor from the land-grant institutions Iowa State, Wisconsin, and Michigan Universities who had been assigned on a survey job for the Chicago-Missouri and St. Paul railroad in the 1880s. It was not a happy meeting of work cultures: "These men seem to think they were out for fun only, and I had to say to them, that it made no difference whether they were from the slums of a city, or a State University; if they remained in my camp they must behave decently." Ainsworth concluded: "the work should be done in my way."[30]

Others agreed. "That the professional schools can make engineers is absurd," wrote one reader of the widely read New York journal the *Engineering News* in 1915, expressing a strong belief in shop-floor knowledge.[31] And even educators had to concede that, "engineers are not made in college." Many agreed that only a "long apprenticeship of practice" marked the true rite of passage into engineering. The engineering student "learns to be an engineer by his post-graduate course in life, where he is rubbed bright by continual practice."[32] Such instruction in the workplace directly challenged the authority sought by engineering educators. The editors of *Engineering and Contracting* of Chicago, for instance, quoted their managing director, the civil engineer and scientific management adept, Halbert P. Gillette (b. 1869), who had argued that a "complete education should give a man habits as well as ideas and training in logic. The habit of going among men, the habit of studying their habits...[are]...certainly not less important than a training in science. I fear they are, however, the very things that few educators have tried to cultivate in their students."[33] The 'habit of going among men' represented a managerial ideal of engineering that carried a heavy political load. It thus linked engineering professionalism closely to management and business.

Educational reformers like Thurston who sought to upgrade engineering training faced a dilemma, however. Their form of engineering knowledge was based on the authority of science rather than on the authority of the workplace where class shaped the relations between management and labor. In the eyes of employers who were supposed to hire their students, academic knowledge did not prepare them for the rough-and-tumble realities of the production floors and building sites. Many engineering educators tried to imitate "the methods and manners of real shop-life" in college shops that housed steam engines, blacksmith tools, foundries and the like. Here hands-on experience could be acquired while preserving academic ideals. The issue did not turn on machines but on men. However well-equipped, the problem with the college shops was that the true confrontation with the attitudes of independent workers and bullying foremen could not be tested. The ability to "handle men" remained the true hallmark of the successful engineer of a management professionalism of engineering. This managerial ideal of engineering balanced precariously between working-class manliness and academic gentility.[34] Work in the laboratory, the shop, and the field, tough jokes, and overalls all added luster to the male rites of passages into the profession. It served not only an educational goal, but also sought to enhance the prestige of engineering education. In these environments, women students were encouraged to take math classes but often excluded from taking shop or field trips to factories mandatory for graduation.

The balancing act between labor and capital, between working-class manliness and female gentility found visual expression in the representational strategies of advocates of engineering education. The young California School of Technology chose as the cover of its new magazine the image of Douglas Tilden's sculpture *Mechanics Fountain*, which celebrated the muscular masculinity of working-class men. The sculpture, which fashioned a fanciful machine correct in mechanical detail (lever, fixed pivot, and pivot link) but with an unworkable design, offered a sanitized vision of the union between capital and labor – the ideal that engineering educators espoused. The image of virile working-class men catered to a thoroughly middle-class audience. It also featured young apprentices dangling dangerously from the lever arm of the punch press. [Figure 12] Consciously or not, the Californian engineering educators promoted an image that was strangely appropriate for the thousands of engineering students who, upon graduation, were desperately seeking entry-level jobs in the tight market of the 1890s.[35]

As engineering educators were aware, many graduates found themselves stuck in drafting departments rather than climbing the promotional ladder towards full-fledged engineering careers. Educators were caught between their own desire for further formalization to meet the standards of academic colleagues of the sciences and the demand from industry for practical training of their personnel. They depended for the most part on the willingness of powerful employers to accept engineering graduates. To circumvent such contests over control and command in the American workplace and the dwindling opportunities in the American West, academic engineers helped their graduates gain practical experience on the building sites of the fast-expanding American empire from Cuba, Panama, Nicaragua to Hawaii and the Philippines in the aftermath of the Spanish-American war. In a debate over the most appropriate engineering education, bridge engineer and advocate Waddell insisted that Spanish rather than French should be taught because "I am fully convinced that the United States will soon dominate the foreign business of Latin-America; and that such a result must come about primarily through

Figure 12.   Academic engineers projecting an athletic, muscular, and working-class but precari-ous manliness on the cover of the first issue of *The California Journal of Technology* 1,1 (February 1903). Apprentices are dangling dangerously from the lever arm of the punch press correct in mechanical detail but with an overall dysfunctional design. Courtesy of University of California Library, Berkeley, CA.

the efforts of American engineers."[36] For many, the emerging empire became the moral gymnasium for young graduates to gain entry into the field through the rites of passages outside the U.S. in the period after 1898.

In the American South, Northern industrialists and philanthropists helped draw racial boundaries around technical expertise during the same period of expansion. Many African-American slaves had received technical training through apprenticeships that had been used to teach slaves the vocational skills needed to keep plantations self-supporting. They worked as skilled stone masons, blacksmiths, and inventors. Funded by Northern industrialists, promoters of the New South movement realized that after the abolition of slavery they needed the scores of freed slaves as a new labor force to industrialize the South and defuse the hostility of white young men to do work tainted with the association with manual labor. Technical education for the freed slaves seemed a logical step after Abolition, but African Americans did not benefit from the allocations of the 1862 Morrill Act, intended to democratize higher education for the sons and daughters of farmers and workers. Instead, a coalition of missionaries, freedmen's bureau's officials, and Northern industrialists helped fund separate institutions for African-Americans including the Hampton Normal and Industrial Institute in Virginia (1872), Atlanta University, and the Tuskegee Normal and Industrial Institute in Alabama (1883) to inculcate a Northern work ethic and industrial skills. Even the revision of the Act in 1890 stipulating that no state would receive federal money if it did not admit people of color to its land-grant colleges failed to solve the problem because it also mandated that the states could provide such an education in "separate but equal" institutions. As a result of the second Morrill Act, sixteen federally supported and separate African-American land-grant institutions and seven state-run African-American normal schools and colleges were established between 1890 and 1915.[37] African-Americans were shunned and sent onto separate educational and employment tracks in the useful arts, but not the applied sciences.

The additional funds for African-Americans provided separate educational paths towards technical work, carefully redrawing the lines of race when many Southern African-Americans started to move to the North in the hope of escaping the growing oppression in the South. This legislation effectively kept African-Americans from joining the ranks of the expanding engineering occupation. In the hardened racism of the 1890s, African-Americans relied on several strategies to negotiate the narrow space left to them. At the Southern black schools, engineering topics were taught under the disguise of the useful arts, while Howard University in Washington DC started a genuine engineering program within its own walls as early as 1912. Individual African-Americans tried to scale the bastion by applying to white Northern engineering schools including Ohio State University, Yale, and MIT. The tactics to train African-Americans for technical fields was at least suc-

cessful enough that by 1927 a newly established engineering organization, the National Technical Association, could boast 125 academically trained members. Even
though they were engineers by training, it is significant they were careful to avoid
the term and instead chose to identify themselves as technicians. Similarly, for African-Americans, the mechanic arts might have been useful and profitable, but
they were not allowed to be termed applied science.[38] The African-American leaders Booker T. Washington and W.E.B. Du Bois disagreed on what strategies to
follow in this atmosphere of hardened racism during the 1890s. The Southern
ex-slave Washington opted for a tactic of camouflage by which he mobilized the
rhetorical positions of the useful arts for his own use. To the younger Harvard-
educated Du Bois, true liberation would succeed only if African-Americans had
the right to be Shakespeare scholars and engineering professionals on the same
terms as whites.

REVITALIZING MALE AUTHORITY THROUGH PROFESSIONALIZATION

At the end of the nineteenth century pressures for inclusion came from all sides: labor, women, and African-American advocates all demanded their place. Through
the language of the medical model of professionalism advocates of engineering recast a male middle-class discourse of technical mastery and control that had also
been part of the patriarchal culture of business model of professionalism. The first
contest of class identity between the advocates of the shopfloor and those of
school-culture socialization had been limited to the establishment engineering.
The second challenge to the reproduction of old patterns of male authority between establishment engineers and the rank-and-file was fought in terms of different models of professionalization. Academic engineers, engineering advocates,
and establishment engineers employed rhetorical strategies and coined new words
to support these efforts. Their border disputes with other workers grew out of a
small elite's efforts to establish professional organizations which reflected a larger
movement of professionalization also witnessed in many other specializations like
law and medicine.[39]

As an occupation, American engineering both resembled and differed from
other new professions because it was a divided house that was never able to establish the classical gatekeeping mechanisms like licensing laws, uniform education,
and credentialling, other professions were able to mobilize. The conflicts over the
true path towards an engineering career came to the fore in a battle over the kind
of professional model engineering would take. As engineering transformed into a
mass occupation and leaders tried to control the process, three professional models
competed for prominence in the first half of the twentieth century. Business lead-

ers and establishment engineers working for private business patronized a professionalism steeped in management ideals of command and control. It had an ideological base in the shopfloor culture of the Eastern industrial establishment and had been reworked to suit the circumstances of industrial capitalism.

By contrast many academic, consulting, and urban engineers supported a professionalism modelled after the medical profession's emphasis on autonomy and ethics that would keep a distance from the business ethics of profitibility. Rank-and-file engineers rejected both management ideals and ethics; instead they focussed on employment services, working conditions, and pay through separate engineering unions and bargaining units within blue-collar unions. In acceptance speeches, professional journals, and anniversary banquets, engineering advocates began to claim infinite expertise in aspects and directions far beyond purely technical knowledge, but they simultaneously expressed the need to establish clear and unambiguous boundaries within an occupation where anybody from the boardroom and the research lab to the drafting department could claim to be engineer. Engineers engaged in different forms of boundary work because they did not succeed in drawing on the classical gatekeeping mechanisms of the other professions. The establishment of the early occupational organizations provided one means for developing the much-needed sense of unity that engineering lacked. These organizations helped to fix firm boundaries at points where they seemed particularly porous. In 1867, the older generation of civil engineers had established the first professional association (ASCE), which sheltered older, wealthier, and conservative members and upheld exclusionary requirements for new members for many years to come. In response to these elite organizations, more inclusionary local organizations that could better deal with the regional working conditions and welcomed young graduates sprung up everywhere in the urban centers of Pittsburgh, Chicago, St. Louis, San Francisco, and many other industrial communities. In San Francisco, the Technical Society of the Pacific Coast served as a focal point for engineers as far as Hawaii, while the Cleveland Engineering Society tended to the local politics of the city's reform mayors. By 1915, one observer estimated that about 20,000 engineers had banded together in local clubs and the same number in national associations. These initiatives were so successful that within thirty years, 200,000 engineers had organized themselves in national societies; another 12,000 belonged to state associations, 27,000 to local ones, and 72,000 joined engineering honor societies and fraternities.[40] Many held multiple memberships, knitting the growing fraternity together that came to carve out an important niche of male middle-class identity, authority, and values. Professional organizations played important roles in setting occupational boundaries on both edges – upper and lower.

Just as America expanded its industries westward and onward into countries in South America and the Pacific and the field of engineering became a mass occupation in the process, engineering advocates left no stone unturned to define and re-strict the field in their struggle for professional standards. In their efforts to clarify the social boundaries of their occupation, they often defined their identity in exclusionary terms. Instead of specifying which skills warranted the title of engi-neer, their definitions centered on those who failed to qualify for certification as an engineer and should therefore be excluded from the field. For example, in support of the new constitution of the American Society of Civil Engineers (ASCE), Herman K. Higgins, who had worked for the railroads for many years in the U.S. and had reached the level of an assistant engineer on the politically tainted, ill-managed, and malaria-plagued Panama Canal project, forcefully argued in 1907 that "the definition of the term engineer should exclude the surveyor per se." Higgins' call for a clear professional boundary came on the heels of a major reorga-nization of the demoralized labor force and engineering staff at his workplace in the Panama Zone after the management structure of the railroads. Leading urban engineer John A. Bensel (b.1863), who had been formally educated at Stevens In-stitute of Technology in the 1880s and practically trained on site in urban building projects such as water, sewage, and tunnel construction in New York City, sec-onded Higgins a few years later. As the then President of the ASCE, Bensel, warned that "in our accomplishments we are not to be measured as skilled artisans but the fact remains that...society at large does so rate us." In like manner, another member of the engineering establishment deplored in 1910 "the loose manner in which the term 'engineer' is applied to men in many departments of work, and of-ten appropriated by those who have no proper claim to be so called." In 1917, the editor of *Engineering and Contracting* in Chicago offered some practical advice on how to curb the "usurpation of this professional title by mechanics," proposing that both railroad officials and unions be persuaded "to drop the use of the word 'engineer' as a designation of locomotive-drivers," and endorsing those who spon-sored licensing laws for engineers. He expected that such legislation would be the most effective instrument in restricting the free use of engineering titles.[41]

Champions of professionalization of the electrical industry took similar rhetor-ical positions aimed at exclusion. In the 1880s, there had been no offense in calling businessmen-engineers like Alexander Graham Bell and Elihu Thomson electri-cians or telegraph operators. To put an end to this practice, the newly established professional organization, the American Institute of Electrical Engineers (AIEE), interfered linguistically with the introduction of the term "electrical engineer" as the field became too crowded. Thomas Lockwood, a long-time telegrapher, ex-plained to the members of the AIEE in 1892, for example, that the term "electrical engineer" had been adopted when a great number of people called themselves elec-tricians and "the word [electrician] fell into some disrepute." He recalled that, "it was necessary to coin another and more euphonious one," and also pointed to the irony that "thus it came about that before we had any institutions for learning in that line, we had electrical engineers."[42]

For similar reasons, leaders in the chemical industry like Arthur D. Little introduced the new term "chemical engineer" to enhance their professional identity. In the case of chemical engineering, this act grew out of deliberate attempts to offset the inadequate pay, low status, and lack of promotion of chemists – many of whom were women who worked in corporate research laboratories. Chemical engineering advocates like Arthur D. Little were determined to distinguish themselves from ill-paid research chemists and analytical chemists. Chemists earned considerably lower wages because their work in chemical research and testing labs was often routine, monotonous, and dangerous. These chemical research jobs offered white men few opportunities for promotion to move up and out of these unpleasant working conditions, although it did provide the few opportunities available to women and African-American scientists. When asked by the Woman's Bureau after the first World War, Little asserted that chemical research offered excellent employment opportunities for women, but he rejected these occupational models for men. To offset the problematic reputation of chemistry work, Little and other founding members of the American Institute of Chemical Engineers (est. 1908) deliberately associated themselves instead with the production process and with male management after Taylorite fashion rather than with industrial research departments or academic science.[43]

Thus, the semantic shift from chemist to chemical engineer established an alignment with management rather than with science work. The shift was not a rhetorical construct alone. It sealed a professional struggle that was in part political. It linked chemical engineering to the patriarchal authority of business and management practices in a time when the chemical industry was consolidating into large corporations. As a consequence of this intense boundary work, chemical engineers could become plant managers rather than lab technicians.

Engineering reformers shared many of the same professional and class ambitions as lawyers and doctors, but there were also important differences between them. If doctors and lawyers served largely unorganized clients, engineers faced the formidable force of powerful political and corporate organizations.[44] Engineers faced the challenge of mastering both the ability of businessmen to offer tangible proof of their trade's worthiness, and the knowledge and skill of the specialized workers whom they sought to manage and on whose cooperation they depended so considerably. Their managerial demeanor acquired a more distant style when they supervised contractors and gangs of common laborers like the Irish diggers of the North American canals, the Italian masons and the Chinese track workers in railroad construction, or African-American convicts in both industries. In this workplace context, the engineers with managerial skills were understood to be male and white, as a matter of course. Increasingly engineers found themselves not merely caught in the middle negotiating a narrow space between labor and capital, but also acted as an active and self-conscious constituent of a middle class in formation. Engineers remained distinct in that they straddled the capitalist fence between the interests of proprietary firms, large corporations, and political machines on the one hand, and skilled and unskilled workers on the other. "The technical man," wrote one engineer in an engineering advocacy journal in 1918, "will join with neither capital nor labor but stand upon professional and correct ethical principles."[45]

This position in the middle turned out to be an increasingly tenuous one. In December 1917 when the U.S. had entered the war and many in the profession began to articulate their worries, the chief engineer of the Indiana Public Service Commission, Otto H. Garman (b. 1880) fashioned the ideal of successful engineers in politically loaded sartorial subtleties. "Let me say it is all right to wear the overalls during the day, but have a different suit of clothes and use fine soap and water before going home in the evening," he advised his engineering colleagues.[46] Educated at land-grant institution Purdue University and employed as an urban engineer connected with the urban management movement that sought to circumvent local politics, Garman projected an ideal world where engineers deftly shuttled back and forth between the world of work and leisure, building site and office, working-class and upper-class, labor and capital, and men and women. By donning the proper garb, engineers could hold these opposing domains together, he believed. Garman belonged to a group of city managers who maintained tenuous relationships with corporate leaders, who sought to wrest control of cities from the working class and the political bosses, and who worked for the kind of issues advocated by the many women's clubs pushing for civic improvements. During the 1910s, this middling position became idealized in the notion that engineers occupied a unique and privileged space between labor and capital.[Figure 13] For patrician and urban engineering reformers like electrical engineer Morris Cooke, the only guarantee of preserving independence in the middle was through the tactic of classical professionalism, with its emphasis on autonomy and ethics, shedding, like the rhetorical strategy of technology, its working-class, classic blue-collar associations. However, this precarious position was rather idealized and not attainable for all. Garman, Higgins, Bensel, Cooke, and other advocates of an engineering professionalism that emphasized ethics and esthetics represented only a minority.[47] [Figure 14]

For rank-and-file engineers the ability to maintain this middle ground involved much more than donning the proper garb. They sought an engineering professionalism less preoccupied with ethics than with bread and butter issues including pay scales, medical insurance, and vacations without being associated with blue-collar

Figure 13.   Idealized engineer in a Veblenesque position of Confidence between Labor and Capital published in the engineering's advocacy journal, *The Professional Engineer*, (1924). Note the archetypical but subtle sartorial distinctions of head and foot gear.

Figure 14.   Idealized male path of promotion from surveyor on the building site to a management position at the corporate office when all was not well in the profession. Reproduced from
The Professional Engineer (January 1923).

unionism. Their professionalism was neither modelled after the business community nor after the medical occupation. Whereas engineers seeking management
professionalism saw themselves primarily as proprietary-managers and the reformers viewed themselves as city managers or academics in public and university employ, rank-and-file engineers came to accept their status as employees.

Worrying about the working and social conditions, rank-and-file mechanical
engineer Norman McLeod (b. 1879), the son of an Irish immigrant father and a
graduate from Cornell with work experience at the large electrical corporations,
expressed the sentiment of scores of others who labored on the lower rungs of the
occupation at the beginning of the century. In the same year that Garman pro-

posed his sartorial politics, McLeod sent a letter to the newly established American Association of Engineers (AAE) to express his frustration with the society's leadership. It might be very well for the AAE to talk about raising ethics and status, he wrote, but instead of blaming engineers for the working conditions, he believed that "The ignorant capitalist...is the place to begin [raising ethics] and not with the engineer." The specter of a medical model of professionalism, with its emphasis on raising standards and ethics, might be a promising tactic for engineers in improving salaries but, McLeod warned, the organization failed to address the more basic need of finding engineers jobs: "Do not continue to advertise engineering students and practical men 'for positions at small salary with the prospect of advancement.'" McLeod, who had been trying to find a job on his own for two years when the U.S. was preparing for war, wryly concluded his letter: "This advancement usually comes when the man is dead."[48] Another letter writer agreed with McLeod's sentiment that the patrician leaders' emphasis on ethics was out of touch with the daily concerns of most engineers at the bottom rung of the profession. He employed a military and engineering metaphor to ventilate his anger and charged: "If you are going to raise the standards of ethics very much higher, you will have to provide it with an oxygen tank so that it does not start frothing at the mouth. What you need is an organization provided with 42-centimeter anti-aircraft guns to puncture the bag of ethics and bring it a little nearer to the earth. Just now we are so full of ethics that there is no room for your economic and social welfare stuff."[49] Thus the engineering rank-and-file had different priorities than their more successful brothers at the top.

Many historians of the profession, however, have preferred to study the concerns of Garman and his peers rather than those of McLeod.[50] They regard Garman's search for status and for professional autonomy as a classic expression of middle-class aspirations resulting from responses to social pressures from above. In this view, the process of professionalization functioned as a mechanism by which engineers attempted to appropriate their status mainly from their social superiors or competitors – a status that could be shown by donning fine business suits and maintaining clean-shaven faces. This interpretation should come as no surprise given the choice of sources to which most scholars have resorted in mapping the terrain. If one thumbs through the pages of professional journals and takes the words of articulate establishment engineers at face value, the expressed anxieties about status can easily be seen in terms of the relationship between engineers and their superiors. To be sure, establishment engineers like Garman dealt in their daily work with a host of businessmen, scientists, and politicians on whose power and favor they depended for employment and recognition. But rank-and-file engineers often felt the pressures from below, as McLeod insisted. The danger of being declassed always presented a threat – a risk that entailed having to replace a white

Figure 15.   Many engineering graduates' final destination was corporate drafting rooms like
this one at the Baltimore and Ohio Railroad in 1899. Courtesy of Division of Engineering and
Industry, National Museum of American History, Smithsonian Institution, Washington, DC.

with a blue collar and facing the peril of falling off the pay scale into the trench.[51]
McLeod's concerns serve as a reminder that middle-class solidarity of American
engineers did not result from contests with politicians, businessmen, and scientists
alone. Most rank-and-file engineers articulated their sense of middle-class entitle-
ment and privilege in a contest with those they encountered on a daily basis in the
workplace, like accountants, chemical analysts, draftsmen, surveyors, electricians,
telegraph operators, common laborers, migrant miners, and skilled mechanics.[52]
Unlike in Britain where engineering was more closely bound to the culture of family businesses or in France where its mathematical and theoretical orientation
linked the profession to the state and to science, in the U.S. boundaries on both
edges – upper and lower – defined the very fabric of American engineering identity. On the lower edge of the occupation, the struggle of rank-and-file engineers
with skilled and unskilled workers – subordinate or competing – equally shaped
their unstable middle-class identity.[53] Thus, a crucial but neglected mechanism of
the engineers' identity also stemmed from the dynamics in the workplace. It
sought to reproduce a distinctly male, white, and middle-class identity for a new
era.

Broken Paternal Promises of Promotion

Despite the considerable sociological and historical literature on the subject of professionalization, younger and rank-and-file members thus felt more concern with personal opportunities for promotion than with professional ethics or ideals of national progress. Rank-and-file engineers like McLeod had little interest in professional autonomy and found their way to the AAE because of its employment services.

The fine lines between engineers and blue-collar but increasingly also white-collar workers came to fashion engineers' distinct American middle-class identity. Between 1910 and 1930 when businesses merged and consolidated, the opportunities for promotions in large corporations dwindled. This thereby jeopardized the paternal promise of promotion that had been essential when engineers belonged to an elite. The vast majority of engineers – young graduates and rank-and-file seniors – found themselves doing routine, monotonous work as surveyors, draftsmen, tracers, copyists, and calculators in drafting departments behind desks, with little hope of advancement, as one young graduate observed at the American Bridge Company in Philadelphia. Like so many other big firms, the company employed thousands of draftsmen who designed, traced, and blueprinted bridges from New York to Uganda. Robert W. Shelmire (b. 1881), an engineering reformer and municipal engineer at the Bureau of Design in Chicago, was very concerned with the possible explosive nature of the issue. He stated the unspeakable when he said: "most of the young men are draftsmen. Even competent engineers also are draftsmen. If the truth were known, one would probably find that the bulk of the entire profession were leaning over drafting tables."[54] [Figures 15 and 16] To be sure some engineers still adhered to the managerial or classical professional ideal and

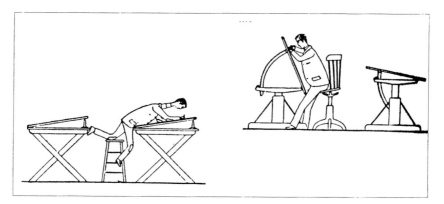

Figure 16.   Drawing illustrating the rough working conditions and long hours of most draftsmen leaning over drafting tables. Reproduced from *The Draftsman* (1901).

worked as proprietors, entrepreneurs, managers, consultants, and professional experts witnesses, but the majority worked in the employ of others as draftsmen, tracers, blue printers, or lab assistants. "The engineer as a rule works for someone else, and is not his own boss unless he is so fortunate as to be able to engage in private practice," as one commentator remarked in 1917.[55] These working conditions were not limited to the field of civil engineering. Few mining engineers worked as operators or had jobs in managerial positions, as the idealized images of popular writers would have it. Many earned their livelihood in claims offices from the anthracite district in southeastern Pennsylvania to the coal and silver mines in the northwest region of Coeur d'Alene in Idaho. The increased division of labor in engineering work meant that young engineers found their upwards path cut off. For instance, about 20 percent of mechanical engineering students who graduated before 1896 eventually became members or partners in a firm. That number rapidly dropped to ten percent in 1904. "It is doubtful if 10 per cent of graduate engineers become engaged in business for themselves, either as consulting engineers or contractors", one contempory asserted.[56] At the same time, the number of graduates who became or remained draftsmen increased. This was the other side of the increased division of labor and segmentation.[57] As the opportunities for entrepreneurial ventures decreased, large corporations began to hire lower middle-class youths and some well-trained middle-class women for low-level jobs, resulting in shifting patterns of mobility.

By the 1910s, the engineering occupation showed serious points of tension between established engineers and the vast majority working in large corporations many of whom were draftsmen working as copyists, tracers, blueprinters or designers. The frictions were not new; they had existed throughout the history of engineering. In earlier times when discontented with the establishment's business-as-usual attitude, individual engineers had refused to pay their dues for professional organizations while mining engineers in Colorado seriously contemplated joining the unions.[58] But for a long time these conflicts never took any organized form. This friction became articulated and organized only in the 1910s and 1920s when engineers faced the increasing consolidation of businesses into large corporations in the aftermath of the first World War.

In the ensuing confrontation, engineering identity became most ideologically explicit in a variety of organizations and also found articulation in the politics of the workplace. Shelmire, learning his engineering skills through correspondence school and practice, warned just after the first World War, "Engineers have been raising themselves as it were by keeping the draftsmen down. Naturally draftsmen have become discontented; hence the prevailing and nationwide unrest in drafting rooms."[59] In the aftermath of the first World War, the discontent of engineers in low and midlevel positions interested in bread-and-butter issues and the criticism of patrician reformers who pursued a classic form of professionalism led to a temporary but powerful alliance both in the American Association of Engineers (AAE) and in its predecessor the Technical League. Under the leadership of patrician reformers C.D. Drayer and F. Newell, the AAE attracted some 22,000 members in 1921, about 15 percent of the profession. It not only pushed for a code of ethics, a major concern of the patrician reformers, but also campaigned for licensing requirements and a salary scale for engineers. In 1909 its predecessor organization, the Technical League, had already launched a direct attack on the limited opportunities for promotion and the exclusionary policies of the fraternity of the American Society of Civil Engineers (ASCE).

Welcoming the establishment of the Technical League in 1909 "with delight," a young engineer from New York City then wrote: "let us no longer disguise our poverty under a mask of dignity." Younger engineers, he asserted, "can expect very little encouragement from the generals of the profession who are mainly responsible for the conditions in which the young engineers find themselves." Most engineers were "herded together as thick as the size of the room will permit with no other accommodations than a rough table or part of it, a stool, and a drawer."

Even if most no longer expected to set up their own firm, it was a true affront to the engineers' self-esteem and manhood to be treated as if they belonged to the class of clerical workers, like bookkeepers or low-level clerks. Worse still, this junior engineer implied that because engineers did not have unions to protect them, the senior engineer in charge "generally exhibits less respect for the individuality of the young engineer than for that of a mechanic who has the manhood to tell him what he should be paid and how many hours he is to work."[60] He acknowledged that although a ceiling prevailed for the relatively high salaries of mechanics, by contrast, the wages of engineers had no roof. But here lay the crux of the matter. The viability of promotion and the ability to move through the ceiling of the wage scales was very much in question. "The proportion of adequately paid engineers to the total number engaged in the profession is so small that, unless he be a genius," the disgruntled and indignant engineer stated, "his opportunities for earning a comfortable salary even at middle age may be considered extremely limited." An editor of a draftsman's journal in Cleveland, Ohio warned that draftsmen who did mental work but whose employers paid and treated them as workmen were most envious of the "better paid union workmen in the machine shop." The editor offered the tactic of drawing linguistic boundaries around the draftsmen's occupation by reserving the term engineer for the "draftsman who is a responsible designer, and whose work is as truly to be ranked with the work of professional men as that of engineers in any other capacity" at the expense of copyists and tracers who "might better their situation by trade union methods." A chief draftsman "is in reality a constructing engineer," pontificated another.[61]

To counter the overcrowding of the engineering labor market, young and formally educated engineers pushed for licensing legislation and succeeded in sponsoring the first law in Wyoming in 1907.[62] The strategy of licensing as a way to limit the access to the labor market had been highly successful with other aspiring professional groups like lawyers and doctors, but as the editor of the liberal *Engineering News* retorted, such comparisons did not apply because "doctors and lawyers are almost invariably independent operators; they neither hire nor are hired by their colleagues. Engineers, on the contrary, in their early years, and sometimes all their life, are in the employ of other engineers, and this brings up the eternally disturbing relation of master and servant." Engineers' direct dependence on the workings of industrial capitalism made them more vulnerable and less protected than lawyers and doctors. Generally sympathetic to the grievances of young engineers, the editor conceded that unionization would be an effective way to safeguard better pay and working conditions, but argued that the solidarity of workers which buttressed the strength of unions was precisely the point where engineers differed. Although unionization might financially benefit engineers in general, he thought it would void the promise to the individual engineer of becoming a manager: "If such a subordinate [junior engineer] ever expected to become a commander, he will hardly join a society the end of which will be the establishment of a lower grade of engineers, where the salaries, to be sure, will be higher than in the lower grades now, but whose members must remain at this slightly elevated level with small hope of advance."[63] At the heart of the self-esteem of engineers, as the young aspiring engineer and the outspoken editor understood it, was an acceptance of low pay and middle-class solidarity in exchange for the promise of promotion and the opportunity to become a commander or a captain of industry.

Thus, the promise of promotion expressed a paternal pledge between junior and senior men of the middle class, albeit one severely tested and frequently broken in the first half of the twentieth century. This broken promise made eminently clear the very tenuous "middling" position in which the majority of junior and rank-and-file engineers found themselves. In an article entitled, "What is Wrong with the Engineering Profession?" a civil engineer assessed the great discontent among engineers in 1915: "The great body of engineers...are on a par with the artisans, and there is small distinction between the ordinary draftsman or field man and the mason and mechanic. This is forcibly brought home by the fact that the wages paid to each are practically the same."[64] For engineers on the periphery of the occupational hierarchy, the prospect of further advancement greatly diminished and caused great anguish. They expressed their anxieties by evoking images of slavery. Such characterization was an exaggeration, to say the least, but it did reflect the false image of mastery held out by the "commanders" of the profession who resided in corporate boardrooms to the bulk of younger and rank-and-file engineers who were leaning over drafting tables.

The evocations of slavery, however overstated, had their bearings on the trend of declining engineering wages. Before the second World War, local markets had depended on a variety of factors including the price of labor, the particular industry, and material resources, and the infrastructure largely determined wage differentials. But even if wages varied according to locale, specialization and industry, on average, wages had been higher throughout the nineteenth century. In the middle decades, civil engineers had been among the highest paid railroad employees. Mining engineers continued to earn high incomes, particularly between 1895 and 1914 – the Golden Age of mining – when American engineers with experience in the mining practices of the western United States were in great demand both at home and abroad from Chili to China. Furthermore, the discovery of precious metals deep below ground in countries like South Africa resulted in a plethora of job opportunities. Chemical engineering salaries fluctuated most often, whereas civil engineering pay tended to be lower but more stable. Regardless of their specialization, young engineering graduates would start their first job on the same salary. At the beginning of their training, civil engineer trainees who worked as rodmen received the same pay as common laborers, while the lower assistant engineers stood to earn slightly more than skilled laborers. But the wage differentials could become quite substantial as the engineers entered their thirties and had worked for about ten years – the very moment young engineers tended to get promoted.[65] If they managed to climb the promotional ladder, successful engineers earned very high salaries. Only a few managed to command such generous incomes, however.

Engineering wages – and this became a sore point for engineers – declined proportionally to other professions with which they competed for social standing. From 1929 to 1954, a period characterized by economic booms and busts, available data indicate that the average pay of engineers declined in relation to doctors' income although it still remained well above that of college teachers. The decline in wages occurred when the profession enjoyed considerable prestige and when public opinion polls regarded engineering as one of the most desirable careers for a young man. At the same time, the engineering establishment continued to lobby for more funds for technical education, arguing that a shortage of technical personnel was near at hand which threatened to stunt national economic growth. The warning of an imminent shortage was not borne out by the continuing decline of engineering wages, however. More importantly, perhaps, their claim symbolized their ongoing push for public recognition and professional prestige instead of a realistic assessment of the nation's needs. In the twentieth century, especially after the second World War, engineering wages decreased more sharply relative to the

Figure 17.   Cartoon "Hang on!" expressing engineers' anxiety of unemployment and sense of precarious position between labor and capital in the post-war depression. Reproduced from *The Professional Engineer*, journal calling for better pay and promotion opportunities during the 1910s and 1920s.

income of other salaried employees.[66] Clearly this was not slavery but to those men who had been promised a place in the middle class it felt that way.

Thus, behind the continuing worries about pay scales lurked a more basic fear of being declassed and demasculinized.[Figure 17] When the federal government decided to recruit women for the war economy, a 1917 editorial on comparative wages singled out the following: "our government is paying typists direct from business college $100.00 per month. They are paying experienced tracers on electrical and mechanical design, $75.00 per month; general draftsmen, $100.00 per month; experienced detailers $125.00; and engineers with training and experience for responsible designing, $150.00." Spokesmen of the chemical industry "blushed" when a 1921 *Herald Tribune* article hinted that chemists might be classed with char women and laborers. The message was clear: the promise of upward mobility would turn into a joke if typists just out of school earned the same or even more than aspiring engineers. On the way to upward mobility from drafts-man to engineer in responsible charge, one had to leave behind the office where the typists spent their days. Similar precarious boundaries had to be maintained between engineers and working-class men, if engineers "had any chance for being raised out of the class of high grade mechanics among whom he now accepts his place." In a more joking fashion, municipal engineer M.Y. Crowdus at the City Engineers' Office in Nashville, Tennessee, vented his anxiety about the wage level of engineers when he offered a rhyme to his engineering fraternity: "It gives my heart a painful wrench/ to know the fellow in the trench/ That I designed, draws in three days/ more than my *weekly* striving pays."[67]

Although money accounted for a large part of the discontent voiced by engi-neers, the issue had a direct bearing on the engineers' self-image and served as a living proof of the viability of upward mobility. Lacking the gatekeeping mecha-nisms of uniform licensing or education, the defining process of engineering took place in contests with other men and women in the workplace, where the space for negotiation continued to narrow. The decline in wages seriously challenged the dignity, self-respect, and expectations of engineers.

While the urban workshops of Philadelphia and the expanding fields of the em-pire of American industrial capitalism had been a moral gymnasium for male so-cialization for men of Taylor's class and an early generation of young graduates of newly established engineering programs, the boundaries between white-collar and blue-collar work grew porous in the large corporate bureaucracies after the first World War. Thus, a dual peril lurked behind attempts to preserve an engineering identity as it became bureaucratized on the one hand, the danger of loss of status, and on the other, the threat to middle-class manliness. The majority of engineers began to acknowledge this threat as the gap narrowed between their wages and those of skilled workers, to whom they considered themselves superior. The dan-ger of being declassed, therefore, presented a frightening prospect to an engineer's sense of identity. Yet when radical reformers put forth the possibility of unioniza-tion as a way out, the idealized image proved a major obstacle. The young engi-neer, who had been so critical of the "commanders" of the engineering field, keenly observed the dilemma: skilled mechanics had unions to protect their man-hood, but engineers lacked such lobbies because they continued to believe in indi-vidual distinction rather than economic solidarity. In effect, the belief in promotion prevented them from joining or forming unions. As one engineer wrote, "there is no halfway landing. An engineer is either a professional engineer or a union man – he can't be both."[68] In the aborted movement of the AAE from 1915 until 1925, engineers like Garman and McLeod had argued over what direction the organization of engineering reform should take. At the end of a period between 1870 to 1920 when engineering became a mass occupation, most American engineers preferred a male professional identity, however symbolic, over conventional, blue-collar unionism. Unlike their British counterparts, they could differentiate themselves from manual workers because of their access to schooling and the promised management positions, and from women engineers, technicians, draftsmen, and chemists who labored under various job titles often masking their high education and training. But that middle-class identity could only be tenuously maintained.

Making Technology a Mask for Disunity

Engineers had an almost knee-jerk aversion against blue-collar unionism, even if during the downturns in the engineering labor market during the 1930s and 1940s their middle-class loyalties were severely tested. By the end of those decades, many engineers had accepted their employee status. New engineering organizations sprung up in the period between the 1930s and the early 1950s. In the 1930s depression, those engineers who were able to find jobs saw their wages decline by a third, while 34 per cent of their colleagues experienced bouts of unemployment. Unemployment hit 20,000 chemists and chemical engineers in New York City. In Cleveland, Ohio, 1,400 of the 4,500 were out of a job; some of them were sent to farmland outside the city to "cultivate truck gardens, raising crops for their own use or for barter."[69] The government-financed construction spree amounted to $6,500,000,000 and was responsible for tunneling mountains, damming rivers, laying aqueducts, digging canals, restoring national landmarks, clearing slums, and building school houses. For engineers who had been clinging to their male middle-class identity finding themselves digging trenches was a true affront that demanded marshalling all their psychological resources in these times of economic bust. One married mechanical engineer with seven children, who dug ditches for the Civil Works Administration in 1935, wrote anonymously of his adjustment to blue-collar work: "The biggest thing to learn is to bring your mind down to the ditch."[70] Nevertheless, he was fortunate to be digging at all. Nora Stanton Barney, a feminist and civil engineer, correctly pointed out in 1933 that the Rooseveltian reforestation, highway, building, and reclamation programs were all closed to qualified women of her generation. Moreover, the National Recovery Board still specified lower minimum wages for women than men while mobilizing men at lower wages far away from their homes, leaving women to head their households.[71]

In response to the economic crisis, independent and self-employed engineers resorted to the classic tactic of licensing that had proved to be so successful for the medical profession in an effort to protect the engineering labor market from crowding. Working outside the large corporations, these largely self-employed engineers of small firms established the National Society of Professional Engineers in 1934. In the period between 1937 and 1957, rank-and-file engineers, many of whom worked in large corporate and federal bureaucracies as technicians, organized into engineering unions. After the passage of the Wagner Act in 1935, industrial unions began to recruit engineers into their ranks. The AFL affiliate American Federation of Technical Engineers and the 1934 established CIO affiliate International Federation of Architects, Engineers, Chemists and Technicians recruited the lowest of the profession like draftsmen. Even so, because the CIO engineering union, like other labor organizations, failed to support the Equal Rights Amendment for women, equal pay for equal work remained an illusion for many women at the lowest rung of the profession. More popular were the exclusive engineering unions united in the umbrella organization the Engineers and Scientists of America, which kept a safe distance from traditional blue-collar unions.[72]

Engineers did not succeed in lending authority to the occupation through licensing controlled by a central state agency like their counterparts in France or in the American medical profession. The authors of the 1930 Wickenden report on engineering education refused to standardize education or support licensing. Instead they opted for an open, heterogeneous, diversified, and stratified educational system, fully aware that not everybody could – nor should – rise to the top. The strong opposition of powerful employers prevented such licensing laws from becoming effective, while the federal government did not push for a centralized agency. As a result, the locus of the defining process of engineering – what it meant to be an engineer – lay neither exclusively in educational institutions nor in license-granting agencies. Instead, the workplace was an important arena where engineering identity, authority, and control were shaped. Unlike continental European models, working experience rather than education represented the most important marker for American engineering identity and authority. Getting a promotion was the crucial marker of an engineer's identity. If, in the course of the nineteenth century, most of his middle-class identity had been shaped in the noisy, dirty, and pedestrian surroundings of the production floors in the Eastern corridor, the railroad beds in the Mid and Farwest, and the mine shafts in California and Colorado, by the twentieth century, his identity was more tenuous as he tried to hold his ground in the basements of drafting departments without falling off the pay scale into the ditch.

The encroachment of industrial corporations increasingly rendered the male professional ideal of autonomy and control an illusion. [Figure 18] The corporate ideal ran counter to a middle-class manly ideal based on individual autonomy, initiative, control, and creativity; it demanded that employees restrain their individualism and give in to the corporate collective. The public-relations department at DuPont Co., in 1951, phrased it thus: "The industrial chemist today stands in

Figure 18.   Dupont advertising of 1922 projecting corporate image of the "lone-wolf inventor" when the company was consolidating into a giant enterprise employing thousands of ill-paid chemists. Courtesy of Hagley Museum and Library, Wilmington, DE.

sharp contrast with the lone-wolf inventor common even a little more than a generation ago. Although the initiative and creative thought of the individual is still of prime importance, the modern researcher is a team player with a strong sense of cooperative effort."[Figure 19]. The remaking of this corporate manliness, coined the "Organization Man" by General Motors's President Alfred Sloan, did not go without a hitch.[73] Because salaried engineers had to contend with pressures from both above and below, relinquishing building sites and production floors to women would have endangered the delicate balance of mastery and middle-class manliness that was increasingly under attack in the growing federal and corporate

Figure 19. Dupont public-relations' department self-conscious 1951 caption to this photograph ran: "The industrial chemist today stands in sharp contrast with the lone-wolf inventor common even a little more than a generation ago. Although the initiative and creative thought of the individual is still of prime importance, the modern researcher is a team player with a strong sense of cooperative effort." Courtesy of Hagley Museum and Library, Wilmington, DE.

bureaucracies. The exclusion of women from the production floor sealed a tacit pact between the fraternity of elite engineers and the rank-and-file engineers. It kept alive the promise, often unfulfilled, that upward mobility was still a viable option for middle-class men in the twentieth century.

By the 1930s engineers' professional status proved to be largely symbolic as engineering elites blocked any effective mechanism of closure or effective professional unions. Business leaders, research scientists, and academic engineers including John Waddell, Arthur Little, Karl Compton, Robert Millikan, Michael Pupin, Charles Kettering, and Frank Jewett continued to upgrade engineering by establishing schools, associations, and journals; they mobilized scientific authority and new funds through such organizations as the National Science Research; and they projected an overarching vision at America's World Fairs. Together with corporations like Dupont and General Motors they pushed the 1933 Chicago and 1939 New York World's Fairs as part of an elaborate public-relations campaign to divert attention away from the discussion about the issue of technological unemployment and the unemployment among engineers.

Engineering might have been a divided house and a segmented occupation, but it was precisely this disunity that prompted engineering advocates to claim the universe for it. They laid claim to the entire universe and needed a catch phrase to articulate their vision of their professional legitimacy in part because of the ambiguity of lacking a specific technical object. The refurbished term "Technology" – now incorporating the discourses on machine aesthetics, academic engineering knowledge, and anthropological notions on civilization – provided that. Establishment engineering advocates defined their domain of expertise as "the exploitation of natural forces."[74] This self-proclaimed mandate was so broad, and the skills needed for it were so varied, that the definition offered little common technical ground for its practitioners. "The true object of engineering is not to create machines...[but]...a mastery in the application of the laws of nature," as John J. Carty (1861-1932) told an audience of his fellow electrical engineers in 1928 at the end of his very successful career in corporate employ. The chief telephone engineer supervising more than 2,000 researchers at the Bell Corporation's research and development department, he went on to say that "man will be liberated, and...the forces of the universe will be employed in his service."[75] He echoed the frequently quoted formulation of Thomas Tredgold (1788-1829), a British engineer of an earlier generation who had defined the field of engineering for the Charter of the British Institution of Civil Engineers as "the art of directing the great sources of power in Nature for the use and convenience of man" – a formulation which fell squarely within the tradition of useful knowledge and borrowed from an ideology of possessive individualism not available to the corporate employees he supervised.[76] To be sure, Carty spoke about man metaphorically, but the singular also suggested a kind of individuality unavailable to most working in the consolidated corporations. After the second World War "Technology" employed as a keyword became an important rhetorical device weaving competing threads together into a corporate fabric when the profession was deeply divided and its main tenets were questioned.

3

Bargaining for the Fraternity

Engineers not only created bridges. They also produced culture in the gray areas of privately printed autobiographies and speeches at birthday parties. While at work on the production floors, building sites, and in laboratories, engineers produced culture as well as goods. They engineered plots that carved out a space in the middle: between labor and capital, Anglo-Saxon culture and ethnic strife, technical reports and literary autobiographies, crass Gilded-Age industrialism and genteel liberal arts. That position in the middle was a tenuous one which they tried to maintain by revitalizing a code of manliness. More than forty civil, mining, and mechanical engineers believed their lives worthwhile enough to write about them.[1] Without apology, they traced the turning points of their lives along a chronology of work and promotion, cataloguing the formation of the most male-dominated occupation of all. Even if all had much to say about their pioneering efforts neither women like Emily Roebling of the Brooklyn Bridge and the first engineering graduate Josephine Zeller nor African-American engineers like successful engineer-entrepreneur Archie Alexander and the industrial researcher Gordon Grady put pen to paper. Autobiographies constituted an approved literary genre for white middle-class men, who mobilized the genre when they faced profound changes in their profession.

Autobiographers related their lives and emotions to their colleagues, addressing the fraternity of their fellow engineers, exhorting young men to follow their example, and excluding blood relatives and women from the narrative. In their male address from one generation to the next, engineers attempted to reproduce a younger generation of men bound by class, gender, and race.[2] Working in different industries covering a wide geographical area, these scribbling engineers might have led diverse lives, but in their autobiographies most employed a similar style of surveillance: tunnel-vision observations, dry technical descriptions, and disembodied prose. This style served a rhetorical strategy for a managerial professional male identity that sought to transcend the field engineers' day-to-day associations with the smell, dirt, and noise of the workers they supervised. They did so by reaching out to their professional brethren across the nation. In the depth of the depression

and debates about technological unemployment, these autobiographies were reworked into stories of success, manliness, civilization, and national identity.

The first of these autobiographical products – often printed privately or endorsed by professional societies – entered the literary scene on the heels of the economic crisis of 1893 that ushered in a period of industrial warfare, corporate
mergers, imperial expansion, and urbanization. The majority of autobiographies
were published during the 1930s depression when the civil building boom had lev-
eled off and America's imperial projects – from the transnational Pacific railroads
in the U.S. to the Panama Canal and the urban infrastructures ranging from Ha-
vana, Manila, and Kyoto – came to a close. More often than not, the engineers
who sat down to write about their lives came from the middle stratum of America's first modern business and government enterprises, notably the railroads and
urban centers. Out in the field, railroad engineers often faced disappointments
about their status and, in writing their autobiographies, worked through, sorted
out, selected, and made sense of the confusing experience of their 'middling' position in capitalist production. On the hissing production floors in the steel industry, mechanical engineers contested powerful skilled workers and their unions,
and in writing an autobiography they claimed a technical expertise over and
against them. In the scaffolded cities, municipal engineers came to see themselves
as the stewards of good government etching out a position between working-class
ward politics and aggressive business communities. They too celebrated their position in the middle.

Written in the decades following the 1893 economic crisis and cultural reorientation to the 1930 depression, the forty-odd autobiographies mark a collective moment of memory that was at once nostalgic and generative.[3] To be sure, they
expressed the characteristic desire of autobiographers to invest in forms of work
that will outlive the self, but, even if ineffectively, they also responded to the new
demands for role models in a vastly altered and rapidly changing profession when
the profession was transformed into a mass occupation.[4] The midlevel engineers
crafted their narratives when American corporations consolidated into large bureaucratic structures, the number of job opportunities for engineers dropped, and
rank-and-file members of the profession began to articulate their discontent with
their seniors. If in the 1890s the authors merely addressed other men within the established engineering fraternity, from the 1910s onwards, the narratives began to
exhort a new generation of young men anxious about career perspectives to emu-
late the lives of their predecessors when the paternal pledges for promotion were
broken. They waxed nostalgic as they described lives and careers no longer open to
the young men they addressed during the depth of the 1930s depression. Collectively they marked a moment when rank-and-file engineers and young engineering graduates were beginning to give voice to their discontent over the increasing

lack of opportunities for promotion and employment. When discussions about technological unemployment threatened to implicate the scientific and engineering communities in the human misery and many rank-and-file engineers found themselves out of a job, a wave of engineering autobiographies appeared on the commercial market. The octogenarian engineers who wrote and published their autobiographies during these pivotal decades spoke to, but not necessarily for, the young engineers or, for that matter, the rank-and-file, who had not been promoted as expected.[5] From a literary point of view, however, they failed as authors because they rarely made themselves the center of their narrative focussing instead on technical details. Yet their literary progeny helped revitalize a middle-class identity for men when the profession was transformed from an elite profession into a mass occupation.

Carving out a Space Between Labor and Capital

The engineers who chose to put their lives in writing often came from the nineteenth-century world of canal, turnpike, and railroad construction, industries driven by cut-throat competition and operating at a relentless pace, which hired thousands of Irish, Chinese, and African-American laborers, who graded, shoveled, chiseled, and blasted their way through mountains, deserts, and swamps.[6] Few specialists who worked in the new science-based corporations like the General Electric Company chose to put their professional lives on paper.[7] Nineteenth-century railroads spearheaded modern management facing powerful railroad workers, who pioneered modern methods of collective bargaining, grievance procedures, and union organization. The railroads were America's first big corporations responsible for producing a major segment of America's middle management. Most aspiring civil engineers gained their initial training along the railroad tracks riding the employment waves from boom to bust in the industry before moving on to other stations in their work life. Their experiences with the volatile railroad industry and the labor conflicts shaped their identities in profound ways. Despite the different work practices, they held in common their experience with America's large-scale operations and first corporations. Engineers constituted the larger part of the emerging modern managerial project.

The Southern civil engineer James M. Searles was one of the first to put pen to paper as the 1893 economic crisis ravished the country. Searles exemplified the post-Civil War wedding between old Southern elites and the northern capitalists who sought to mobilize the freed slaves for an industrial base for the New South. Trained with the U.S. Coastal Survey, he quickly moved beyond his first stint with the railroads after deciding that the low pay and the back-breaking labor of a chainman logging a 30 pound transit through the dusty fields was beneath his worth. Instead, he picked the date of the beginning of his career with an appointment as a manager of the levee work in the Mississippi Delta "as it was then and there I was first clothes with the responsibilities attendant on a commanding position." Searles sought to justify himself, to promote the expertise of the engineering fraternity, and to convince Congress to allocate funds for the Mississippi river. "I propose, in writing of the life and times of a civil engineer," Searles explained in 1893, "to speak of my experiences and thereby, perhaps, be instrumental in guiding my professional brethren amid the rocks and shoals which are to be met in the current life of real practical engineering." He hoped that Congress "may be convinced that the Mississippi river problem can be best, most economically, and permanently solved, by outlets, levees, and reservoirs."[8] He also mapped his involvement in many building projects, including the drainage of Cat Island, levee construction down in the Mississippi Delta, urban developments in Nebraska, and railroad building in Kansas, Kentucky, and Alabama – all of which, however, left him bitter and disappointed. Although he had a sense of belonging to the engineering fraternity and maintained various management positions as resident chief engineer, Searles never joined the newly established American Society of Civil Engineers. He belonged to the engineering rank-and-file working as a field engineer near construction sites, but never held a policy position at company headquarters that would have enabled him to influence or direct the plans of construction in any significant fashion. Searles, the son of an impoverished Southern family whose "stomach had been aristocratically trained," had no capital to invest in the projects he believed to be technically sound and commercially profitable. His opposition to raising the levees on the Mississippi River placed him at odds with the established political opinion and the financial interests of the planters, but his minority opinion signalled his allegiance to an emerging understanding of engineers' special claim to technical knowledge.[9]

Searles introduced his narrative with a promise to tell the reader all: how engineers had been cheated out of rewards for technical solutions rightfully theirs that had been appropriated by those "who have been negative factors in the working-out process." The outcome courted disaster because "arrogance can blind the eyes of capital." By the time he was through with his narrative, however, Searles decided "to go back on the promise [to tell all because] I should be obliged to mention names – for the owners of which, in their personal capacity, I have high respect – and whatever may be my estimate of their professional worth, is a matter of no concern to the general public." He felt bound by the "ethical code of the fraternity" that demanded full loyalty and reluctant to divulge any information about his employers. In his loyal attempt at self-restraint, however, he still struggled for self-control and continued to write, "I will, however, say this much, that the proper adaptability of means at hand never fails of legitimately corresponding results."[10] On first reading, Searles' opaque phrase seemed not to tell much at all. It contained no action and no players.

Searles' depersonalized prose is actually very revealing. His reluctance to name names and his choice of disembodied language expresses the narrow space he and his fellow engineers occupied professionally between capital and labor. He enjoyed the owners socially but he mistrusted them professionally and felt vastly superior to the Irish, African-American, and convict laborers he managed on the railroad tracks and the levee construction, even though their employee status resembled his own. Searles defended the engineering fraternity instead of publicizing the capitalists' malpractices he had witnessed, but he considered "it a duty I owe to the many comrades who have faithfully toiled along the hard road of 'out door' Engineering, and to the profession at large, to protest against the money estimate of their services." Before he was done telling, however, he made his allegation and posed his vexing question: "Why don't these small salaried fellows [engineers] climb to the top rung of the ladder?" Having asked no one in particular, he promptly answered his own question: "because money and influential circumstances, uncoupled with true merit, *push* so many up that ladder that the worthy ones can only *look* up."[11] In this closing accusation lay the foundation of a powerful argument for merit and expert knowledge that other engineers and professional organizations would develop in the following decades. Promoting expert knowledge offered a way out of the precarious position he held as a "small salaried fellow" in facing the "arrogance of capital."

Most engineers ranked among the first salaried middle managers and experienced the contradictions of capitalist production as professional workers. Occupying an ambiguous space between capital and labor, engineers epitomized this class conflict. Like wage earners, they faced seasonal unemployment, wage pressures, capital and political control, but like managers, they were also in a position of authority over other wage earners and felt a sense of loyalty to their employers.[12] Searles could put an exact price on his own engineering labor and determine the profit it generated by taking on his employers' perspective, but he also articulated the powerlessness of his position as he described how helpless he felt after the Ship Island Company was sold and merged with the Huntington and Wilson Railroad without his prior knowledge. In expressing his discovery of the merger, he drew on the image of slave labor – unquestionably the most horrifying image available to the son of a family of planters. He recalled his anger over the loss of control of the project as a consultant and chief engineer in these strong words: "*I found myself sold*" [his emphasis], and he concluded, "I have thus written in justice to myself."[13] Through writing his autobiography he bargained for his rightful place. He left no other trace in the written historical record.

The writing of an autobiography remained a singular act for many of these field engineers, whose lives would otherwise have been untraceable. The majority had never written books before embarking on their autobiographies. This is not to say that engineers shunned writing altogether. On the contrary, the writing of logs, diaries, account books, reports, specifications, and contracts formed an integral part of the life of an engineer. Moreover, his daily work required the writing of proposals and of technical reports for public interest groups, investors, and employers. An engineer who advised against the feasibility of a project also lost an employment opportunity; but one who wrote himself into a project often became responsible for its execution and success.[14] Doubtless Searles's appeal to Congress for the building of an integrated system of levees, outlets, and reservoirs for the Mississippi river in 1893 showed his interest in the technical issues in the debate, but the autobiography also amounted to a job application for the new engineering jobs should the federal government decide on financing such a large-scale undertaking. Through writing an engineer bargained for his livelihood.

Searles, raised with Southern disdain for manual labor and hands-on training so celebrated by sons of the North, composed an autobiography uniquely his own. As a middle-level manager, Searles occupied a half-way position in the capitalist production and the political system of the New South. He recast his identity from a planter's son to a civil engineer in the aftermath of the Civil War.[15] Several other Southerners also insisted on a depoliticized, neutral sounding life story.[16] Searles's autobiography contained themes common to the dozens of other engineers – both from the South and the North – who followed him in the decades to come. By writing an autobiography, he and his colleagues pried open a space for the engineering fraternity. He expressed the horror engineers felt at the narrow space for which they had to negotiate as employees in one of America's earliest corporate industries. In his efforts both to tell and to suppress his story, Searles unveiled a male engineering identity in the making at the end of the nineteenth century.

WRITING A WORLD WITHOUT WORKERS

The engineering autobiographies constituted a curious literary genre. They drew on a variety of literary forms including biographical sketches, portraits, memoirs, and obituaries that appeared in engineering journals alongside numerous articles discussing the proper definition of an engineer. From a literary point of view, however, these were 'failed' autobiographies, as the engineers spun their narratives along elaborate descriptions of engineering projects that read like technical reports rather than narrations of the self.

No doubt the engineers' professed unwillingness to reveal their lives followed a nineteenth-century autobiographical convention: they often claimed to have begun writing only after insistent cajoling from family, friends and – more significantly perhaps – out of a sense of loyalty to the engineering fraternity.[17] But engineers had many legitimate reasons for their reluctance. In the words of one self-reflective autobiographer, engineers possessed poor penmanship. Accustomed to rendering three-dimensional forms on paper and trained to thinking visually and spatially, engineers disliked expressing their thoughts in writing. Another engineer recalled how he was "once told that an engineer who wrote a book was not an engineer," reflecting a hardened sentiment of anti-intellectualism in engineering that is widely prevalent in engineering self-representations in America.[18] Fully aware of the engineers' poor penmanship, academic engineers, organized in the Society for the Promotion of Engineering Education in 1891, worried about "the disgraceful ignorance of their native language on the part of a large majority of our students and alumni," in the words of leading bridge engineer John A.L. Waddell.[19] Pushing for the inclusion of liberal education in the curriculum of engineers proved to be one of the many tactics they employed in raising the status of the profession. Beyond the lack of liberal preparation and literary practice, aspiring autobiographers had other worries. Engineers ranked among the first professionals to confront corporations that often forbade them to divulge any information about the projects they had worked on – and engineers had no intention of questioning the rules of an entrepreneurial class to whom they owed their sustenance and loyalties, as Searles made clear.

As middle managers engineers felt responsible for the company and internalized their employers' demand for secrecy in their dealings with investors, competitors, and subordinates. Investors demanded the utmost secrecy from surveyors and engineers during the projecting and surveying stages, when large sums of money and potential profits were at stake. "Secrecy as well as haste is frequently a good qualification for an engineer," counselled field engineer Danforth H. Ainsworth (1828-1904) with a long career behind him in railroad employ in the wide expanse of the Northwest across the Mississippi river. As a company man, Ainsworth had struggled with establishing his authority among villagers charging high prices for food, shelter, and water, with maintaining control over local contractors, foremen, teamsters, and rodmen unwilling to obey his orders, and with keeping land speculators at bay. "It sometimes calls for a good deal of policy on the part of the engineer to avoid publicity," he asserted.[20] In a similar vein, one mining engineer told a young aspiring man to be tacit when handling different competing commercial interests. He advised, "Be as Cold Blooded and as Unenthusiastic as a Clam."[21] This strict sense of loyalty could weigh so heavily on an engineer that he might refrain from publishing his autobiography altogether. Sworn to secrecy for the sake of a

company's success and profits, civil and mining engineers confronted the contradictions of corporate loyalty and the circumscribed space allowed for individual
promotion and expression that was so critical to middle-class identity and male
prowess. They were in the position of entrepreneurs without the capital to see a
project through; they also worked as employees with full responsibility without
any of the entrepreneur's power and freedom to negotiate. Mid-level managerial
engineers like Searles and Ainsworth were wedded so much to the ideals of capitalist production that they wore their discretion like badges of professional pride; but
their faith in the business ideology was severely tested when many projects came to
naught because of local politics, failed capital, mismanagement, or conflicting visions.[22] Their autobiographies expressed both their need and their reluctance to
bear witness. For these mid-level engineers to publicize themselves at all was both
remarkable and painful.[23]

   Ainsworth showed how excruciating it could be for engineers to reveal themselves in a direct and active voice. He had participated in the official engineering
fraternity as a member since 1886, but as a field engineer with his boots deep in the
mud supervising construction sites and organizing labor camps along projected
railroads west of the Mississipi expanding into Illinois, Iowa, Missouri, Minnesota, the Dakotas, Nebraska, and Colorado, Ainsworth could not afford to participate in the ASCE's day-to-day discussions held far away in the smoke-filled rooms
on the East Coast. He had benefitted from the boom in railroad construction during the 1870s, enjoyed managerial command, and could call himself a certified
member of the fraternity, but Ainsworth nevertheless occupied the lower stratum
of the managerial hierarchy. When he wrote his autobiography at the end of his career, he was finding it increasingly difficult to get work at his usual rate, feeling the
brunt of the collapse of the railroad construction industry. Recalling his precarious
situation as a civil engineer in railroad location and construction, Ainsworth referred to himself in the third person, using a passive voice: "His position," he said
of his own situation, "was never a source of pride to him, and was anomalous to say
the least. He was the company's disbursing agent, and deemed himself the ever accessible dog to be kicked whenever anything went wrong." Ainsworth used the
same stylistic devices in describing his anger at the State of New York for losing his
job on the Erie Canal: "Possibly one knows he is hurt when trampled upon, even if
he has no corns. There is quite a distinct recollection," he said of himself, "that a
tongue was not under complete control, and perhaps the interest of fellow-
sufferers were not for the time considered. The State of New York still owes what
was then justly claimed."[24] The occupational demands for secrecy shaped the
mode of expression of Ainsworth and other engineers. They were not only cautious with information, but also practiced an economy of expression, prized terse-
ness as a professional asset, and employed disembodied prose.

The engineers replicated their occupational stance of inspection, supervision, surveying and surveillance that objectified land, environs and people.[25] Like the style of the surveyor who takes meticulous notes on the profiles of the land and the inhabitants with little personal involvement, civil engineer and surveyor Henry Root phrased his surveying work with the Central Pacific Railroad in a 1921 recollection thus: "On July 5, 1866, I went by train to Secrettown and by stage from there paying $6.20 stage fare and stopping at Dutch Flat for dinner. I lived in a log cabin on the east side of Crystal Lake from this time until December 23, 1866, when I moved with McCloud's party to camp 41 to go into winter quarters to give lines and grades in Tunnels 3, 4, 5 and to work on estimates."[26] Root was not alone in writing short sentences, using simple syntax, and emphasizing materials, dimensions, and production instead of dwelling on the labor practices on the building sites, an angle of repose in their home lives, or the magnificant landscapes they were about to transform. Nor was Ainsworth's use of a third-person narrative and passive voice exceptional. Through these conventions engineers established their distance, objectivity, and veracity.

The journalist Bess Demaree observed the disembodied landscape that engineers constructed, too. 'Like many other engineers,' she wrote, John D. Littlepage, an American mining engineer who worked in the gold industry for the Soviet Gold Trust Company between 1927 and 1937, "can describe a mine with complete clarity and accuracy, but seldom bothers to describe a person or scene. And he thinks, more often than not, in terms of production."[27] This tendency also struck Cecile Hulse Matschat (1895-1976), the wife of an engineer, who highlighted the gendered differences between herself and her husband in observing the South-American landscape. "My husband was so absorbed with the odd little engine, whose chief function seemed to be to hook the cars to the cable," she wrote of a joint trip, "that he paid no attention to the magnificent view. But I refused to be enthusiastic over what looked to me like a glorified elevator. I thought the scenery was stupendous."[28] Landscape was an asset to the professional lives of both husband and wife, but, as a botanist, Matschat looked at the landscape as a source of knowledge, beauty, and culinary resources: her gaze was expansive, peopled, and colorful. By contrast, her husband Louis could not see beyond the engineering aspects of the site.

The terse narrative style of engineers also reflected the changes in communications within the emerging corporations. As employees in modern enterprises, engineers regularly used new forms of communication like the telegraph and telephone to transmit messages across the great distances separating the office from the construction site. Civil exchanges of information between Eastern financiers sitting in board rooms and managers far away in the field mattered little to the railroad corporations when haste and the high operating costs of telegraph and telephone were involved. Field engineer Ainsworth recalled how the corporations showed "little disposition to be communicative" to management engineers like himself and pushed employees for economy of expression. Once when he wired a message to ask: 'Will the company build 100 miles this year?' The answer promptly returned was 'No,' and when he sent another asking: 'Will 60 miles be built this year?' he got the same negative reply followed by curt instruction: 'You will be censured here if sixty miles are not built this year.' Of course, the corporations' demand for economical use of language also embedded a form of command and control in the emerging corporations. In a 1913 report in which officials of the chemical corporation Dupont Company ordered the use of more efficient language for internal corporate communication, they asked managers to replace a phrase like 'we are in receipt of a request from St. Louis office for' with 'St. Louis asks for.' Similarly, they urged managers to replace a request which began with 'we would be glad to have you make an examination and advise,' with the formula 'please examine and report.'[29] As managing the overload of information became a serious concern of growing public and private corporations, engineers helped to shape a new, more economic language of command and control without subtle negotiations. They confronted and participated in the historical shift away from face-to-face communication toward increasingly mediated and indirect forms of exchange between superiors and subordinates.

The engineers thus wrote their autobiographies along the lines of the dry narrative style of technical reports and the form of their daily communications more than along any literary conventions. The intense descriptive gaze of Matschat's husband, Ainsworth, and other engineers also turned the readers' attention away from the surveyor himself.[30] From a literary point of view, these were 'failed' autobiographies that read like technical reports rather than like narrations of the self. Poor liberal education, corporate control, secrecy, information overload, a desire to transcend local circumstance, and a claim of technical expertise all shaped the engineers' characteristic writing style: dry technical details, short sentences, economic modes of expression, and disembodied prose. They spoke over and across building sites, beyond the horizon to imagined communities of other engineers. But the disembodied prose also went along with an embodiment of gender.

BUILDING THE ENGINEERING FAMILY WITHOUT WOMEN

Engineers might have been – contrary to autobiographical conventions – slow to position themselves at the center of their own narratives, but they were rarely self-effacing when discussing their dealings with men they considered their social peers or workers whose skills and organization posed a threat to their identities and the knowledge they were claiming for themselves. At these two points – when discribing fraternal feelings or contests with competing workers – critical ruptures appear in their otherwise disembodied narratives.

Many engineers portrayed themselves as independent builders and producers, thus obscuring the intricate labor relations involved and the bitter industrial disputes that left deep imprints during these decades. They appropriated the physical activities of building and execution characteristic of the skilled workers they supervised while, in reality, they merely designed, organized, and supervised those workers who actually shovelled dirt, laid bricks, worked lathes, and lifted steelplates. In their writing, they often asserted their technical authority over skilled workers. John Fritz (1822-1913), a mill foreman who climbed up the managerial ladder in the Pennsylvanian steel industry, had supervised the construction of a steel mill complex by bringing under one roof Bessemer converters that could burn carbon out of molten iron by blowing cold air through it under high pressure producing steel more cheaply and introducing rolling mills that shaped red-hot iron into sheets. He appropriated the language of building, implying that he had built a Bessemer plant and a rolling mill with his own bare hands. "I built a good substantial stone building," he said of a construction job that required a great deal of time and involved squads of masons, bricklayers, and common laborers.[31]

Fritz and other mechanical engineers often appropriated the skills of other workers in their narratives, if only to [re]claim a position of authority over them. Electrical and mechanical engineer William LeRoy Emmet (1859-1941), raised in a family left in financial ruins by the after shocks of the Civil War and with little interest in formal learning, gained on-the-job training in the hierarchical work culture of the American navy and electrical corporations like Westing House and General Electric. In one of his first jobs for the Sprague Illuminating Company as a troubleshooter on electric streetcars in 1888, Emmet organized a group of Italians into a team of mechanics to assist him in the overhaul of 120 electrical motors for the Railroad Company in Allegheny City. In his recollection fifty years later, Emmet appropriated the work and skills of the Italian mechanics to highlight his own contributions in the development of an insulating material of varnished cambric, a material he would later develop in the shops of General Electric Co. in Chicago, New York, and Schenectady. Describing his work for the electrical company Sprague in the 1880s, he wrote: "I stripped all the wire from the one hundred and twenty new motors and replaced it in a very different way. I completed one batch of motors before I dismantled another...I made the insulation of varnished embric with no shellac...I also rebuilt all the controllers making radical changes." He then continued: "I also built new trolleys of greatly improved design and sold them to the Railway Company." Only at the point where a contest of skill erupted between himself and the mechanic whom he had ordered to improve a swivelling trolley, however, does he interrupt his narrative of creation, switching back and forth between a language of making and a language of supervision: "I built them," – he wrote of the trolleys in terms of creation and then corrected himself in terms of supervision – "or had them built...in a little machine shop run by a young fellow named R.D. Nuttall." Although he frankly acknowledged the collective process of innovation when he spoke of his peers at General Electric labs in Schenectady, he disavowed this kind of teamwork when it came to discussing the work of the skilled Italian mechanics.[32]

Searles, too, showed how slippery and problematic the language of singular creation could be when he recounted his story. "I was ordered to reconstruct [the bridge]," Searles wrote of one job he supervised, "and had but little difficulty in stopping the rush of water that had destroyed the bridge." He then significantly added, "as I was well supplied with sand-bags and Negro cavalrymen..." Catching himself using the language of building rather than supervision, he self-consciously interrupts his narrative for a clarification. The clarification of the labor process briefly restored the hierarchy of command and control, but inadvertently exposed the racial distance between himself and the African-American laborers. Ridiculing the abolitionists' fraternal love as effete, Searles added parenthetically, "for fear some supersensitive, tender-hearted commiserator of the colored brother might think that I dumped him in the crevasse hole instead of the sandbags, it should be remembered that the cotton of which the bags were made were worth more...than the negroes."[33] In presenting his managing job as an act of physical building, his slip of the pen opened up the problematic race and class relations particular to the South. Painfully aware of his Northern allies, he sought to cover the tension with an intervention of sarcasm, but the damage was done. His two identities, one as a professional engineer and the other as a white Southerner, were in conflict. In his correction, Searles tried to reestablish the proper order of supervision and mastery so vital to the white men of Brady's New South. Searles's narration might have been particularly his own, but his erasure of workers or his self-conscious identification with command, control, and whiteness was far from atypical.

Engineers also suspended their disembodied prose when writing about men of their own social class and often described them as family members. Some authors conventionally began their life stories as a genealogical record, only to mold it into a resume of their professional life.[34] Even when the authors gave genealogical reasons for writing their autobiographies, close relatives like parents and siblings barely figured in the narrative. More often than not they excised spouses and children – including sons – altogether from the genealogy and story line.

If engineers seldom included family members in their life stories or displayed a sense of intimacy, they often wove their narratives around emotional passages involving other men of their social class or their mentors, whom they did treat as family members. Onward Bates (1850-1936), an establishment civil engineer with little formal education, had received his training in the fields of the railroad bridge construction during the Gilded Age before moving to the urban frontier of Chicago to become a contractor. As an octogenarian looking back on his years as Charles Shaler Smith's apprentice, Bates noted in 1933, "I have always regarded him as my professional father."[35] Alfred West Gilbert (1816-1900), who later became a municipal engineer in Cincinnati's waterworks and sewersterms when the city expanded as a result of the shipping business along the Ohio river, recalled a feeling of kinship within the community of engineers on his first surveying mission in the hills of Pennsylvania, but he restricted the sense of family to social equals. Like so many other engineers, Gilbert conveniently excluded their day-to-day dealings with foreign-born common laborers responsible for clearing the path and setting up camp.[36]

Sometimes the fraternal feelings between social equals would go so far as to exclude blood brothers even when they worked side by side. John Fritz, whose brothers George (1828-1873) and William (1841-1884) worked with him in the steel industry at Bethlehem Steel mills in Pennsylvania and at a rolling mill at Chattanooga, Tennessee, failed to portray either brother in a particularly intimate or familial fashion, except for one passage where, ironically, he argued the importance of fraternal love among engineers over and against that of blood relatives. With pathos, Fritz recalled in 1912 the early history of the Bessemer process when his colleagues Alexander Holley, Robert Hunt, William Jones and his brother George would frequently come to Bethlehem to discuss their work in implementing the new steel making process in the plant that produced rails and armor-plate: "We did not meet as diplomats...but we met as a band of loving brother engineers trained by arduous experience, young, able, energetic, and determined to make a success. I doubt if ever five natural brothers were more loyal to each other than the five brother engineers above named."[37]

Urban engineer Onward Bates most explicitly showed how literally this male world could supplant patriarchal family life when he described his engineering work in terms of love and marriage. Fondly recalling the emotions he felt the day he was offered a job with the Edge Moor Company in Wilmington, a firm that specialized in bridge construction: "All I could think of at the moment was what the lady said when she received a proposal of marriage, 'This is very sudden.' I wired Mr. Whittemore: 'I accept Mr. Smith's proposal,'" the man he had earlier described as "my professional father." Bates failed to regard marriage with the same affection, however. He likened marriage to an engineering specification and contract: "We learned that matrimony is not a joke, but a most serious matter, the most important of all contracts, lasting throughout the lives of both parties... These specifications almost warrant one to avoid marriage as too great a risk to be

attempted...”[18] Bates and other engineers erased women, and their wives in particular, from their narratives altogether despite the fact that wives, daughters, and
other relatives lived with them in isolation in the engineering camps and often
acted as the midwives of many engineers' autobiographical offspring in their roles
as collaborators, writers, typists, and sponsors.[39]

The erasure of women's presence and the celebration of the homo-social world
elicited commentary from the women who observed them. In her own autobiography, professional botanist and geographer Cecile Hulse Matschat presented herself wryly, yet seriously, as the dutiful wife of an engineer during the time she spent
in South America, where she had followed her husband on his job. She articulated
the erotic but threatening undercurrent of the fraternal feelings that ran between
her husband and his South-American engineering assistant. Matschat opened her
narrative with a vignette of her husband's assistant, Matéo, and closed it with his
tragic death on the job.

With great detail and much pathos, she described their mutual resentment and
competition for her husband's attention and love. “They greeted each other like
old friends,” she recalled the moment when she first met Matéo as she arrived as a
young bride in South America with her husband. “No one could mistake the
warmth of feeling between them. But there was nothing of affection in the beady
black eyes that swept *me* from head to foot.” Matschat considered her husband's
assistant physically “monstrous,” and as she told the story, each disliked and resented the other not only physically but – only barely hidden under the surface –
sexually. After her husband had introduced her to Matéo as the new bride, the assistant exclaimed: “*That* your woman?” His tone expressed complete disbelief.
“Holy cow's blood!” The gaucho shook his head in disgust. “Damn skinny
woman!” Cecile Matschat wrote of the incident, “I *thought* of a lot of things to say,
but didn't. I *felt* like a rag without a bone and a hank of hair.” She felt betrayed and
abandoned by her husband's failure to perform his duty to speak out for her. She
then continued to describe her assessment of Matéo's capacity for love and sex,
echoing a racial stereotype American whites invoked when talking about African
American men. “I feel certain, now, that Matéo never loved a woman...He boasted
three women, not wives, and to him they were merely goods or chattels, fit only to
work and to perform their natural functions.” And in support of her claim, she asserted in conclusion that Matéo's women merely “awaited the infrequent visits of
their master and the resultant visits of the stork.”[40] Her husband's failure to speak
up for her particularly offended Cecile Hulse Matschat because she expected his
heterosexual and Anglo-Saxon loyalty in this far-away corner of America's expanding industrial capitalism. Matschat's sense of duty to follow her husband to South
America and abandon her career as a professional was predicated on her expectation of the priority of heterosexual love and Anglo-Saxon solidarity. The fraterniz-

ing between her husband and his Latin-American assistant challenged her authority as a white North American woman. Matéo's position in the engineering hierarchy placed him in the same social class as the Matschats, but his ethnicity called for other manners, she believed. The assistant's skills might have been essential to her husband's engineering work and survival but they were not crucial to her, or so, at least, she had assumed. To Matschat, her husband's declining fortunes and mood swings following Matéo's death showed how dependent he had been on his assistant for his livelihood. In the manner she framed her autobiography, Matschat made her resentment of her husband's mixed loyalties quite visceral.

Matschat had not been the first to explore the homo-social world of engineering where the fraternal feelings posed a threat to the women married to them. The sense of loyalty that these corporate men felt towards each other and had internalized vis-à-vis their employers directly competed with the women whom they married or the women who were making professional claims for themselves. Professional women writers including Mary Hallock Foote, Anna Chapin Ray, L. Frances, Elizabeth Foote, and Willa Cather gratefully explored the theme of engineers who were married to the job at the risk of losing the women they were supposed to wed. As we will see, women writers engineered their plots differently than their male colleagues to counter these professional models. As these professional women made clear they resented the way management engineers like Bates substituted marital and blood relations for corporate families. They objected to the construction of communities of men as a world without women; a world where male friendships were marriages and marriages engineering specifications.

APPROPRIATING THE WORKER'S BODY

Civil engineers involved in canal, bridge, and railroad construction maintained a great social distance from the common laborers whom they supervised. The working camps of the diggers of the North American canal system, for example, were characterized by sport, fighting, boozing, and other contests of virile strength that helped to develop a proletarian sense of virility closely associated with manual labor and physical prowess. Keeping a social distance created its own difficulties. Civil engineer James Worrall (1816-1885) might have disliked the pressure to drink with the laborers with whom he worked, yet he also noted that it was impossible to remove himself from the ritual if he wanted to retain their respect for his genteel manliness and authority. He called Benjamin Franklin's famous abstention from alcohol an isolated case, saying that "[the] men who followed his exemplary of life were called eccentric and individual – a kind of prigs...The growing youth of a people naturally prefer to be like the average manly character, and that character is not to be priggish nor exclusive."[41]

Most of his scribbling colleagues ignored the travails of the Irish, Scottish, and French-Canadian canallers, who dug the ditches with shovel and barrow. They removed from their stories the Chinese graders who shoveled, chiseled, blasted and bored their way through mountain ranges and deserts, the Irish workers who grabbed the rails with tongs, guided them over rollers and put them in place, the laborers who started the spikes, secured the fishplates and tightened the bolts, the track levelers who lifted ties and shoveled dirt, or the thousands of tampers who finished the track-laying work with shovels and iron bars.[42] The Southerner Isham Randolph (1848-1920) chronicled his railroad employ of seventeen years from his apprenticeship in clearing the way with axes to his arrival as management engineer before settling in the booming city of Chicago, where he helped plot the railroad's round houses, shops, terminals, and freight houses and then moved to the urban frontier in the Chicago of the Progressive Era. For the next fourteen years he prospered as an urban engineer, gaining local fame for his design of the city's water system – the Chicago Drainage Canal that changed the direction of the Chicago River so that its waters would flow into the Mississippi instead of into Lake Michigan – an experience that landed him a job as a consultant on the Panama Canal later in life. But Randolph never mentioned the thousands of African-American, Polish, and Irish quarry men and canalers who were driven hard to shovel forty miles of sand and solid rock for the much acclaimed Chicago Drainage Canal, nor did he dwell on the events of 1893 when these workers went on a wildcat strike to demand higher wages.[43] Again and again, civil engineers skipped over the building sites and industrial struggles in their stories, and instead fixed their gaze on technical details and reached out to engineering communities across the land. [Figure 20]

The social distance between mechanical engineers and mechanics was more complicated. In the iron and steel industries, mechanical engineers faced the formidable power and skill of tightly organized Welsh and English industrial craftsmen, whose bond was forged by pride of skill and an ethical code of mutualism. In many industrial workshops, skilled master mechanics had been at the center of the enterprise, whose repository of technical knowledge had often been the driving force behind mechanical innovation. These well-organized workers had developed craft-union solidarity as a strategy against the fallout from the 1873 panic and increasing management encroachments on the organization of their work in the heated years during the 1880s. In terms of ethnicity and skill, the Welsh and English mechanics claimed a more intimate link to the social class of mechanical engineers to which Fritz belonged as the son of a German immigrant than the largely unorganized African-American, Chinese, and convict laborers that Searles,

Figure 20.   Plagued by political intrigue, dire working conditions, and outbreaks of malaria, the Panama Canal was built by a work force of 17,000 and organized on the management principles of railroad construction. Here allegorically represented as the work of one man: a modern day Hercules, forcing apart the Culebra Cut to create the Panama Canal. Official poster for the 1915 San Francisco Panama Pacific International Exposition by Perham W. Nahl.

Ainsworth, and Worrall faced in their supervisory duties as civil engineers. If common laborers were safely removed from nearly all narratives of civil engineers, by contrast, the bodies of iron rollers, puddlers, heaters, and mechanics filled the pages of *The Autobiography of John Fritz*. Mechanical engineers like Fritz participated in the rhetorical strategies that attempted to invigorate middle-class notions of manliness and civilization through a focused attention on working-class men.[44]

Engineer-manager John Fritz could look back on a truly prominent career in the steel industry that had been a creation of the railroad and war industries. He was best known for his invention of the three-high rail mill that saved the arm-breaking labor needed to form square bars of iron into finished rails. As superintendent of the Bethlehem Steel mills, he had helped organize the largest plant of its kind in the U.S. in 1860. His life, as he portrayed it in 1912, had been so successful because of his superior technical knowledge even if workers and higher management alike had opposed and reprimanded his bold moves. Most civil engineers who wrote their autobiographies did so on their own, but Fritz's autobiography was almost a collective act of self confident captains of industry who found him the ideal corporate man.

The professional association of mechanical engineers, The American Society of Mechanical Engineers, actively shaped, endorsed, and published Fritz's *Autobiography* as the life of a man who represented the ideal of the mechanical-engineering establishment. His autobiography articulated the growing confidence and aggressiveness of the northeastern industrial establishment, shaped a managerial engineering identity, and cast it as a simple master mechanic ("iron master") who stood by his workers, purging the difficulties that gave rise to it. This was a remarkable feat indeed. The endorsement came when both inside and outside the association tensions rose over the true path toward engineering knowledge. The challenge came from both craft unions who fought the encroachments of management and from a growing band of engineering educators like Thurston who sought to link mechanical engineering with science and its prestige. In marketing a male managerial middle-class identity for engineers, the association played an active role in sponsoring the autobiography of Fritz and other book-length biographies of mechanical engineers including Walter Clark, Fred Colvin, Edward Hewitt, Embury Hitchcock, and Howard Pedrick during the 1920s and 1930s. Fritz's autobiography was published by William H. Wiley (b. 1842), a school-trained mining engineer at Rensselaer Polytechnic Institute and the Columbia School of Mines, a loyal member and treasurer of the organization of mechanical engineers. Wiley and other leading publishers of technical literature like McGraw-Hill steadfastly supported the leadership of industry and management engineers, playing a prominent cultural role in efforts to build their identities through their publications.[45] Fritz's *Autobiography* well suited the purposes of the emerging professionalism of mechanical engineers in their claim to a special kind of technical knowledge which labor unions were challenging during the decades of intense industrial struggle.

Fritz's narrative showed how among mechanical engineers pressures from below were resolved in a celebration of a shop-floor manliness when it was no longer viable. This celebration of the male working-class body went together with the valorization of the bodies of middle-class athletes, a disempowerment of African-American, Native American bodies, and an erasure of women's presence altogether. As the son of a farmer-millwright, Fritz had acquired an early familiarity with the machinery of the cotton mills in Pennsylvania's rural districts where his father earned a living with maintenance work. This was as good a technical education as any young man could hope for. Later, as a supervisor of a large iron and steel operation, he had depended on the tacit knowledge of boilermakers, rollers, and puddlers at the Pennsylvania industrial machine shops of Norristown (1846-1849), Safe Harbor (1849), Cambria (1854-1860), and Bethlehem Steel (1860-1892). Fritz's life exemplified the trial-and-error era of industrial capitalism, when skilled machinists and millwrights (like his father) improved and tinkered with new machines, materials, and metallurgical processes. From a technical point of view young Fritz had not been the *tabula rasa* he and his colleagues made him out to be decades later in 1912. The ASME promoted John Fritz's autobiography as the narrative of a self-made man who had learned by doing. This portrait was quite self-serving because formal education was generally uncommon during the early nineteenth century, only to become retrospectively an issue when school-trained engineers began to overwhelm traditional upper-class engineer-proprietors in their claim for true engineering skills. Fritz's autobiography merged three historical and politically important moments into one powerful argument. The book included reports of the ostentatious birthday parties which the ASME staged in Fritz's honor in 1892 and 1902 – years that turned out to be politically significant.

In the late summer of 1892, the captains, managers, and engineers of the steel industries gathered to celebrate Fritz's seventieth birthday at the Opera House in Bethlehem. After the guests sat through a sumptuous banquet, sipping their coffee, puffing their cigars, engaging in rounds of "merry jest...as friends were recognized up or down the tables...amid a babel of sounds mingled with bursts of uncontrollable laughter", the fraternity not only toasted the story of a self-made man and his manly individualism, but also put the man they had come to celebrate to a mock trial. [Figure 21] Through the practical joke, they charged Fritz with two offenses, accusing him first of having "misled the public into the belief that he was an engineer, and an iron and steelmaker" and, secondly, of having "disturbed the peace." Fritz, so the indictment read, had "changed beyond all recognition the old time peaceful hamlet of Bethlehem." If it had not been for Fritz, "Bethlehem would

Figure 21.   U.S. steel dinner in 1901 celebrating corporate organization similar to Fritz's birthday dinner parties eulogizing his manly corporate character. Courtesy of Carnegie Library of
Pittsburgh (neg. A-146).

have remained to this day the quiet place [it was and] the waving grain would still
be bending to the summer breeze over lands now..," so the jesters charged. The la-
mented loss of a golden pastoral past common in the writings of many European
and American writers acquired a particular meaning for engineers. Many engineers
presented their participation in engineering as contributions to the march of progress and as an illustration of their own self-improvement, but coming, as many did,
from an agricultural background, the authors often expressed ambivalence about
the pastoral world their very profession had helped to destroy. "The indictment
showed," the fraternity charged, "how the prisoner [Fritz] growing up, turned his
back ... on the old farms, sought out a country blacksmith and machineshop,
where he thumped his fingers, greased his clothes, and grew black in the face,
thinking he was becoming an engineer."[46] The creation of a professional autobiography attempted to recapture a past that engineers had helped to destroy, and
served to memorialize a past made obsolete by the very careers they described.[47]

   This concern with a loss of the pastoral past also displaced the contest between
skilled steel workers and managers – a conflict that was on everybody's mind in

September 1892. Three hundred miles away from the celebration at Bethlehem's Opera House, the ongoing and bitter strike at the Carnegie Steel Company in Homestead just outside Pittsburgh in Pennsylvania divided the community if not the country. At the beginning of that summer the conflict had seemed to center merely on pay, but as the months wore on the workers' demands, strategies, and arguments focused on workers' control over the organization of their work. At the heart lay a classic contest over knowledge and skill between workers and management.[48] In this highly charged political atmosphere and amidst a deepening economic crisis, the ASME establishment put John Fritz on mock trial during a banquet given in his honor just as he had retired from a long career as the quintessential manager and corporate man. The invited guests at his table included steel barons like Andrew Carnegie, Henry C. Frick, and Abram S. Hewitt, all of whom were involved in the decisive labor contest at the Homestead steel works. Against the backdrop of the bitter labor-management conflicts at Homestead, the guests found their practical joke so "merry" that they were overcome with "uncontrollable laughter."[49] With a nod to the labor battles being fought in the courts, the organizers accused Fritz of the pretense of possessing technical knowledge and calling himself an engineer; in the opinion of the ASME establishment, the Homestead steel strikers showed an exaggerated confidence in their technical knowledge, as if their skills could not and would not be replaced. Fritz's second offense against Bethlehem's pastoral peace went beyond the inroads the steel industry had made into the agricultural world. A depoliticized narrative of the lost Golden Age thus allowed the ASME establishment to work through their role in, and anxiety over, the transformation of labor relations. Male gendered language obscured and resolved the existing contests between managers and skilled workers in the steel industry. Fritz and the ASME appropriated the skill and manhood of the workers they supervised in language and images: "he thumped his fingers, greased his clothes, and grew black in the face." The accusation of having posed as an engineer playfully referred to Fritz's lack of formal education, increasingly demanded by engineering educators, but in the politically charged air of that summer, it also played on the professional skills engineers claimed vis-à-vis mechanics and management on America's production floors.

The male-gendered language of calluses, dirt, and sweat became more politically coherent in the narrative of 1902 – in the decade characterized by the bitter strikes at the Homestead steelworks, the Coeur d'Alene silvermines, Pullman and Chicago railroads that proved to be decisive in the changing labor-management relations when government intervened in these industrial conflicts on behalf of the owners and managers. This time the mechanical engineering establishment gathered to celebrate John Fritz's eightieth birthday and chose the Waldorf Astoria in New York City as their setting rather than the local Opera House in Bethlehem;

this was a sign of the growing confidence of ASME members in their national importance, and a reflection of their exclusive orientation towards the financial centers of the East Coast. Several guests sitting at Fritz's table had come to despise
organized labor – the product of their conflicts with unions in the decades after the
1873 national depression, the Haymarket scare, and the Homestead strike. Rossiter
W. Raymond (1840-1918), president of the American Institute of Mining Engineers (AIME), an influential writer for his profession and an arch opponent of unions, pitched the manliness of John Fritz against that of craft unionism embodied
in his enemy John Mitchell (1870-1919), president of the United Mine Workers of
America (UMW). Rossiter had developed his vitriolic antagonism against labor
unions when he was working as a consulting engineer for iron manufacturer Cooper, Hewitt & Company during the 1880s; an antagonism he articulated more
publicly and bluntly during the Homestead strike in the early 1890s. At Fritz's
birthday party, Rossiter ridiculed the UMW's struggles for recognition and pay.
"You will not find a great many pages about raising engineers. You will not find
one single plan for shortening a day's work or diminishing the quantity of labor
that an honest man gives for his labor (applause). You will find that Society
[AIME] recognizing individual manhood." Closing his rousing speech, he asserted
that the American manhood of individualism had been divinely inspired.[50]

Abram S. Hewitt (1822-1903) further gendered Raymond's definition of a middle-class manhood. A steel baron and ex-mayor of New York, he had been the
president of the Cambria Iron Works in Johnstown, Pennsylvania, where he had
hired Fritz as a superintendent during the 1850s, and had been Rossiter's boss during the bitter labor disputes in the later part of his life. Hewitt celebrated the fraternal feelings that had blossomed in spite of – and that had also counterbalanced –
the competition that lay at the heart of their relationship. "At times of competitive
struggle, the friendship which has existed between us has never in the slightest
been disturbed," he insisted of his loyal employee. "This happy experience is due,
doubtless, to the amiable traits of Mr. Fritz's nature, which, with all its masculine
energy, is tempered with the sweetness of the gentler sex," Hewitt added in a nod
upwards to the balcony where their wives were segregated in the Angels Gallery
and entertained with a watered-down version of the celebration that went on
downstairs.[51] To the ASME business community, John Fritz's life came explicitly
to represent acquisitive individualism. And in a classic reinstatement of the ideology of the separate spheres, the ruthless competition between husbands was translated into fierce male love, but held in check by the sweet silence of their wives in
the Angels Gallery upstairs. Through its narrative, the fraternity thus framed John
Fritz's life as a paragon of American business and individual manhood, omitting
the struggle that was part of it. For the steel barons and their associates who had
been embroiled in the changing labor relations in the steel industry, Fritz's engi-

neering management professionalism became their ideal vehicle. To emphasize it, they instituted the John Fritz gold medal in 1902.

John Fritz allowed the members of the ASME to shape his life story, but his autobiography – written 10 years later in 1912 – also diverged in subtle and significant ways from their construction. In the third narrative layer of the book, John Fritz recounted his life as a defense of his engineering knowledge, competence, and expertise while carefully stressing his good feelings and relations both with the workers and upper management. Following the tradition set by other engineers, Fritz told his life story through elaborate descriptions of the technical problems he had faced with the staff at the Norristown, Cambria, Safe Harbor, and Bethlehem Steel plants and of his solutions to them. He described conflicts with skilled workers, foremen, formally trained engineers, managers, boards of directors, and bankers alike, many of whom had attended his birthday parties – conflicts from which he emerged victorious over his skeptical, ignorant, and recalcitrant opponents because of his technical expertise. If the ASME defended an ideal manhood of individualism, Fritz's own definition of ideal manhood was to be "a man among men," or projected back into his youth, "a boy among boys" – a theme common among engineers often echoed in the pages of engineering magazines.[52]

The masculine romanticization of the shop floor conveniently disguised the fundamental changes that had taken place and the tensions that gave rise to it. The engineer's ability to withstand workers' jokes and pranks, and his skill in speaking the salty language of the shop floor – coarse to the ear of the intended readers of his autobiography and to his own middle-class sensibilities, but vital for gaining and maintaining his credibility with and authority over the workers – had been forms of initiation into the male world of industry. Fritz insisted, "I do not now forget the laboring man, and especially the able, brave, and noble men who loyally stood by me in times of severest trials...who were ever ready to face any hardship or danger...To these kind and loyal men much credit is due for success as I have attained."[53] In his autobiographical tale of 1912, Fritz sidestepped the tensions between the skilled workers and managers and the industrial warfare that had raged around him: "I wish, also, to give credit to the brave and noble workmen...All that needed to be said was 'Come, boys,' but never 'Go, boys',...too much credit cannot be given to these fearless and energetic men for the marvelous progress that has been made in the manufacture of iron and steel in this country."[54] His insistence on the skilled workers' high level of confidence doubtless represented Fritz's feelings and own social background. It might even have been a subtle criticism of the ruthlessness of the steel barons, but it also masked the sharply contested changes that had occurred in the relations between management and skilled workers at Bethlehem Steel, Cambria Iron Works, and elsewhere in the steel industry throughout his lifetime. Fritz had been celebrated for his invention of the

Figure 22.   This 1840s "hook tool" was used by the skilled machinist who tucked its wooden handle under his armpit to steady the cutting edge against the rapidly spinning work piece in the lathe. Presented by John Fritz in his Presidential speech to the American Society of Mechanical Engineers in 1896, it functioned as a relic of the mechanists' skills, a material trophy of corporate progress, and a fetish of working-class manliness. Courtesy of Division of Engineering and Industry, National Museum of American History, Smithsonian Institution, Washington, DC. (neg. 45210-A).

three-high rail mill that saved the arm-breaking labor needed to form square bars of iron into finished rails. Yet Fritz was also intimately involved and directly implicated in the transformation of another aspect of the labor process: he led the introduction and improvement of the Bessemer Steel process that eliminated the job of the puddler, who constituted the aristocracy of labor and the vocal part of the industrial craft unions' movement.[5] In his 1896 Presidential speech Fritz presented the hook tool as a token of his own and the steel industry's youth to the American Society of Mechanical Engineers. It was also a relic of the machinists' skills, a material trophy of corporate progress, and a fetish of working-class manliness. Shop-floor manliness might have been viable in Philadelphia's machine shops of the 1840s the hooktool represented, but in the late 1890s this was no longer a reality and largely symbolic.[Figure 22] The development of industrialization paralleled Fritz's rise to prominence in iron and steel manufacturing – an industry that had been the major engine for the railroad and war industries. It would become an icon and metaphor for industrialization itself – symbolized at the World's Fairs and sanitized in the presentation of clean machines.

  From engineering's basement sounded dissonant literary voices, however. In ill-lit drafting rooms a few steps below the level of the sidewalk, thousands of draftsmen worried about the height of their stools, the fumes, dust, and noise coming from boiler rooms and foundries, or the treatment they received from their supervisors. The young graduates from the engineering programs, correspondence schools, and evening classes cared little about maintaining managerial modes of manliness. They had come to expect their position in life as employees of

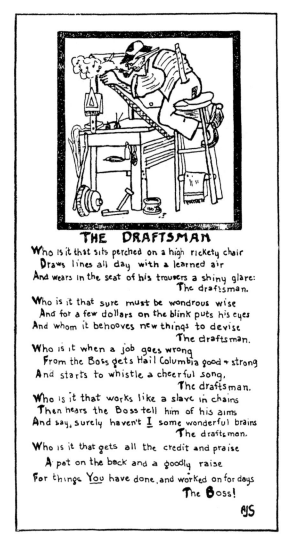

THE DRAFTSMAN

Who is it that sits perched on a high rickety chair
Draws lines all day with a learned air
And wears in the seat of his trousers a shiny glare:
    The draftsman.

Who is it that sure must be wondrous wise
And for a few dollars on the blink puts his eyes
And whom it behooves new things to devise
    The draftsman.

Who is it when a job goes wrong
From the Boss gets Hail Columbia good + strong
And starts to whistle a cheerful song,
    The draftsman.

Who is it that works like a slave in chains
Then hears the Boss tell him of his aims
And say, surely haven't I some wonderful brains
    The draftsman.

Who is it that gets all the credit and praise
A pat on the back and a goodly raise
For things You have done, and worked on for days
    The Boss!

AJS

Figure 23.   Sarcastic poem published in the rank-and-file engineering journal *The Draftsman* (Philadelphia) in 1928 showing that the paternal pledge and promise of promotion to management position was not readily available to most laboring in the basement of the profession.

such large corporations as the American Bridge or the Dupont companies. Through organizations including the Technical League, the American Association of Engineers, and the International Federation of Technical Engineers', Architects' and Draftsmen's Union they began to articulate their demands for better working conditions and pay. When in 1919 the air filled with talk of democracy and revolution, engineering reformer Robert Shelmire articulated the sense of class betrayal many aspiring engineers stuck in drafting departments felt when the paternal pledge of upward mobility was broken. He questioned the refusal of engineering organizations like the ASME to furnish employment services. He asked rhetorically if there was any excuse for failing to help "the younger men who are growing up in the profession and will succeed them. ... The exploitation of young engineer-draftsmen constitutes a most shameful chapter of the history of engineering... Much of the responsibility for the draftsman's predicament is placed on the older engineers. The engineers, the men who style themselves *the* profession,... who have tried the draftsman out of the engineering profession."[56] Instead of composing autobiographies, these draftsmen wrote poems or circulated the verses written by writers like Rudyard Kipling. Poetry counselled the scores of sons from lower-middle and immigrant classes, and some African-Americans and women on how to keep their dignity, helped release the tensions suffered in the drafting departments, and ventriloquized the anxieties at the bottom of the engineering ladder. From the basements, draftsmen approvingly read lines as: "Tomorrow, we will sweep all fears away/Tomorrow – we'll be dead/Go, fool – and play."[57] They counselled on how to maintain dignity in the basement "Don't be afraid of the dirt/That comes from the furnace of the file/Each man who is great... Has worked in the dirt for awhile." They ventilated their anger about bosses. "Who is it that gets all the credit and praise/A pat on the back and a goodly raise/For things *YOU* have done, and worked on for days/The BOSS."[58] [Figure 23]

(Re)Making the History of Engineering

Numerous autobiographies chronicled the many conflicts which civil engineers faced and internalized during the second half of the nineteenth and the early twentieth centuries, and many were, as Searles had made so painfully clear, neither success stories nor narratives of autonomous, self-directed men, but rather bore witness to the narrow space for negotiation which mid-level engineers occupied. This sense of failure and bewilderment shaped the way many authors structured their narratives. Engineering autobiographies, however imperfect they may be as literary works, thus often expressed the confusing experiences faced by individual engineers from the middle strata of corporations. They established their own literary genre characterized by detailed technical descriptions, disembodied prose, and the erasure of workers' knowledge. Above and across the ditches, tracks, and canals, they reached out to the imagined communities elsewhere through their autobiographies. In them, engineers displayed their own technical knowledge, seeking to stake out an area of knowledge and skill entirely their own, while celebrating male bonds. Writing an autobiography became a personal and collective way of reworking past experiences in order to understand, interpret, and recast the present

had become deeply divided. Between the politically sig-
1912, the engineering establishment revitalized Fritz's life
-made man who had pulled himself up by his own boot-
in a language of shopfloor manliness that was dislodged
context."

tion of Fritz's autobiography, the ideal of the self-made
a coherent narrative, a dominant strategy, and market-
eering autobiographies published by commercial presses
became codified as a belief in initiative, aggressiveness,
efulness. Commercial publishers and sponsors in the
' lives as clear expressions of rugged individualism. But
individualism was more a product and a marketing de-
on of the time recalled in the autobiographies. "To me,"
dge in his 1936 foreword to the autobiography of Sam-
1e] has always represented what is best in rugged Amer-
ew him to exaggerate a statement."[60] Likewise, Alfred
west Gilbert's son thought his father's autobiography exemplified "the simple
story of a quiet, modest man, with no claims to either greatness or wealth." He endorsed its publication in 1934, almost fifty years after it had been written, because
"[i]t is good in this distracted time to pause and give thought to those of an earlier
day who served their country...for its development." As if to counter the
Rooseveltian version of state socialism and calls of solidarity, he continued to stress
the importance of individualism as a source of inspiration during a time of economic hardship: "we must depend upon individual character to sustain the nation
which our fathers so laboriously built."[61] In part because of the growing power of
the large corporations, another civil engineer, Paul Starret (1866-1957), who supervised the construction of several of the classic temples erected on the Columbian
Exposition's midway for Chicagoan architects Burnham and Root, stated in 1938
with an air of nostalgia that the era of individualism had already passed: "There
will always be individualism but the era of Fricks and Stillmans and Blacks is of the
past, and with it the builders of that era they are like those vanishing Indians of my
childhood."[62] The autobiographies appeared on the literary scene as the daily press,
weekly magazines, and popular fiction devoted considerable attention to engineers
and their work, thus enhancing and enshrining the profession of engineering. One
establishment engineer noted with pleasure in 1936 that "it is gratifying to the engineers to know that year by year the great importance of the profession is realized
by the general public..."[63] This recognition was perhaps best symbolized by the
election of Herbert Hoover, a mining engineer by training, profession, and political conviction, to the Presidency in 1927, but contrasted with the experience of the
rank and file.

If the authors wrote about the nineteenth century, their autobiographies were also important products of the early twentieth century as they expressed the epoch's taste, tendency, or need for memorializing. Their production and marketing took place during a period which coincided with the passing of the Golden Age of civil, mining, and mechanical engineering. Taken together, the writing and publication of the four dozen autobiographies constitute a significant narrative production and a professional and collective moment of memorializing. During the 1930s, the genre became more firmly established as presses like Wiley and Sons and McGraw-Hill – major publishers for technical, vocational, and engineering schools and active participants in the campaign against the debate over technological employment – publicized the stories of small heroes in the spirit espoused by Ogburn, Giedion, and Usher, who sought to write a new social and anonymous history.[64] The spate of commercially promoted autobiographies was also a defense against the demand of some social scientists that engineers and scientists share responsibility for the economic crisis of the 1930s; they came when the project of building an empire was in decline, corporations consolidated further, and the Federal government was investing heavily in major building through Roosevelt's programs of public-work administration. "It is good in this distracted time to pause and give thought to those of an earlier day who served their country and labored to lay a sure foundation for its development. With all our laws, restrictions, new inventions, and so-called improvements, we must depend upon individual character to sustain the nation which our engineers so laboriously built," justified a widow of an engineer upon the publication of her husband's autobiography, which he had written some forty years earlier.[65] The autobiographies were responses to the growing worries over technological employment during the 1930s and the call for responsibility of engineers and scientists. If their autobiographies had not been commercially viable in earlier decades, during the 1930s commercial presses began to cater to a market that had come to believe that engineers and scientists rather than politicians and statesmen were the small heroes of America and the true shapers of history. These publishers helped name and shape a new area of expertise and knowledge that could for the first time be designated as technology. This history of engineering would be cast as the history of technology.

4

(De)Constructing Male Professional Bridges

Late Victorian male writers of romance and modern artists of the visual arts began to build professional bridges between themselves and engineers. Over the span of two and half decades from 1890 to the first World War, male commercial writers staged the engineer as a male cultural hero. A decade later, the modern avant-garde followed their popular writing brothers by aestheticizing the visual language of machinery and by inscribing machines as explicitly male symbols. Together these two professional groups shaped the symbols of technology. Engineers and machines became the markers of modern manliness but not without protests from women professionals.

If in the autobiographies of male engineers women were removed from the set, in popular engineering novels they occupied center stage. Women played a crucial role in lending engineers their virility. Authors introduced the heroine often to counterpoint or criticize the world of engineering work. No doubt because of literary convention, authors of fiction inserted women as lovers, mothers, and daughters into the engineers' world of work and threw them into sharp relief with male engineers. Within the pages of their own autobiographies, by contrast, engineers described their work as an affair between men only and replaced women and their family of blood relatives by a family of engineers, proving their manhood through their struggles with other men including owners, fellow engineers, and laborers. But in the world of fiction, engineers won their badge of manhood through their associations with women; here the men of fiction acquired their distinctiveness through the women's presence and prodding.

There was another major difference between writers and engineers, however. Male engineers placed their autobiographies in the tradition of vocational literature, which eagerly solicited the attention of young aspiring men, while the writers of popular fiction catered to the preferences of a mass audience dominated by female readers. "It is said that the success of a book," wrote a father-and-son team of writers in the foreword to their engineering novel *Web of Steel,* "depends upon women; that women buy, read, discuss, and promote a novel, and if the book has no appeal to women it is forever doomed." Slyly, the authors, one a popular writer of romance, the other a civil engineer, congratulated themselves for having defied women's dominance in the literary market and "at least proved themselves men of courage, the publishers likewise, for it cannot be too insistently set forth that this is primarily a book...for men, about men, and written by men."¹ As a concession to the female market, the authors constructed their story about men "around the eternal feminine whom the authors have striven to make as feminine and charming, as appealing and delightful, as their large experience with the other sex permits and warrants!" The story line of their engineering novel revolved around the collapse of a bridge and the love between father and son, showing how the demands of engineering work and male honor were incompatible with women's demand for family and community.

The authors of *Web of Steel* articulated a decade-long war of words between female and male authors who had been reflecting on the astonishing rise of both critically and commercially successful women of letters throughout the middle and late nineteenth century. In the winter of 1872, a famous conversation between Mark Twain and Charles Warner and their wives generated the classic answer to the nineteenth-century cultural wars of gender when the husbands questioned the state of popular fiction dominated by such succesful female writers as Harriet Beecher Stowe and Louisa May Alcott. The wives challenged their husband writers to compose a better tale. In answer to the challenge, they published their best-selling *The Gilded Age: A Tale of Today* (1873), a novel about the male industrial world of scoundrels and schemers that would give the era its name.

Recasting this struggle in new terms for the modern age, male and female artists minted the engineer as cultural hero between the economic downturn of the 1890s until after the first World War. Artists often portrayed engineers as visionaries and artists who stood for design, imagination, and leadership, rather than mere execution.[Figure 24] This portrayal fell squarely in line with the eagerness of engineering advocates to show that engineering works were in fact works of art, in an effort to obtain professional recognition from cultural elites.² The portrayal of the engineer as an artist depicted him as a visionary free agent who stood above the mere concern of making money and offered engineers an appealing professional role model.

But the mutual mirroring between engineers and artists reflected a particular male affinity and infatuation, perhaps even a male middle-class alignment of sorts between the male authors' search for social and economic status and the engineers' quest for cultural authority. The gender convergence of the artist-as-engineer and the engineer-as-artist infused both the engineering profession and the writing profession with manly qualities of independence, control, and physical vigor at a time when both professions were increasingly incorporated into modern, bureaucratic, and mass institutions at the turn of the century. Women novelists and artists, however, cared to disagree and devised their own narrative and visual strategies. If

Figure 24.   Engineers as male professional role model for middle-class boys and popular authors stressing the outdoors. Book cover for *The Young Engineers in Mexico* as part of a juvenile series on engineers, 1913.

men ruled in the engineering profession, male authors believed that women dominated the arts. The popularity of the engineers as a protagonist helped to recast that contest in new terms. The manliness of the engineer bridged the professional aspirations of male writing and engineering professionals. Modernist artists also increasingly projected themselves as engineers; they rejected academic traditions and the arduous apprenticeship system of the studio, celebrating instead design over craft. Avant-garde visual artists similarly adopted the machine's aethetics as badges of male versions of modernism to air out the Victorian parlors associated with genteel women. For them, Victorianism represented comfortable and over-stuffed female parlors where women wrote and read in leisure. They adopted engineering aesthetics and devised an identity to counter it. The representation of outdoorsy male engineers, the protagonists of a great many popular writers, served as their agents to air out the stuffiness of the nineteenth-century parlor.

Scribbling Men Design Engineers

Promoters of an engineering professionalism looked for cultural stamps of approval by enlisting writers to enhance their prestige. "The soldier has long been the hero *par excellence* of the writer of romance; the statesman, lawyer, physician, and minister have received their share of glory, and even the business man has not been overlooked...But what of the engineer and his work?" queried Edgar A. Van Deusen in the *Professional Engineer,* home journal of revolting engineers, in 1922. At 38, Van Deusen (b. 1884) could claim an impressive heritage to the Hudson Dutch and to the British engineering profession, and a sound engineering education, but he had been employed in the lower ranks of the profession as chief draftsman with various corporations for almost ten years. After reviewing a number of popular writers' treatment of engineers published in the previous decades, Van Deusen advised that such literature should serve "to give the public a clearer conception of the value to the community of the engineer and his work...Both the profession and the public would owe the authors a debt of gratitude."[3] Van Deusen's call for literary recognition expressed the engineers' more general sense of cultural neglect and public indifference to their professional claims and echoed an earlier plea of J. H. Prior, chief engineer of the Illinois Public Utilities's board, for raising the "social, economic and cultural status" of engineers.[4] Engineering patricians like Prior and Van Deusen canvassed for financial reward and social status but also sought to increase engineers' cultural capital. To engineering advocates, reading popular novels and poetry was more than a leisure activity: it signalled a potential asset that could increase engineers' much needed cultural capital.[5]

Male writers of popular fiction fought their own battles with women authors and with a predominantly female audience. To male authors the "feminization" of the writing profession threatened their professional prestige or what Hawthorne had called a "damned mob of scribbling women" as early as 1855 when he witnessed the stellar rise of writers like Harriet Beecher Stowe and Mary Elizabeth Barrett Browning in the literary marketplace.[6] The engineers' search for cultural authority paralleled the male professional writers' own quest for independence and financial reward. The creation of the figure of the engineer bridged these mutual male professional aspirations. In the period from the Columbian Exposition to the first World War, prior to Van Deusen's rhetorical question, at least 20 novels and short stories appeared on the literary scene in which writers of romance cast the engineer as a white middle-class hero. Published during a relatively short period in the decades preceding the first World War, novels like *Soldiers of Fortune, The Bridge Builders,* and *Still Jim* enjoyed immense popularity at the time of publication, but have long since been forgotten. John Fox Jr.'s *The Trail of the Lonesome*

Pine (1908), Harold Bell Wright's *The Winning of Barbara Worth* (1911), and Zane Grey's *The U.P. Trail* (1918) ranked among the number one bestsellers in the history of American publishing, each selling over a million copies.[7]

The engineer of genteel fiction was a ruling class figure, a visionary, and a leader who was portrayed as an ideal professional: "social, civil, and stable."[8] But this managerial professional ideal was also distinctly male and middle class, and did not include women. In his ruggedness, the engineer became the male successor to the cowboy, the embodiment of physical vigor and control, who conquered and mastered female nature; he symbolized the romantic loner who roamed the country in an attempt to escape the weakening influence of civilization associated with female values. In the portrayal of these male professional writers, the wellspring of the engineer's true manhood was his body. Yet he was also a professional who mapped and civilized the American West, and embodied the middle-class ideal of *The Strenuous Life*.[9] Most fictional engineers were civil or mining engineers who supervised the construction of bridges, railroads, and dams, or the extraction of precious ores for the expanding American empire; few, if any, dealt with mechanical, chemical, or electrical problems. The fictional engineer was muscular and did not hesitate to get his hands dirty, but he was never confused with the laborers he supervised. He was a leader and a loyal company man married to his job, and as such he became the premier male and middle-class role model of work during the 1910s for which the patrician Van Deusen longed.

One engineering story first serialized in *McClure's* and illustrated by Charles Dana Gibson before being published as a book was Richard Harding Davis's *Soldiers of Fortune* (1897); it was emblematic for the new genre. It tells of a mining engineer, Robert Clay, who exploits the Valencia Mining Company for American absentee owners in a small South American country called Olancha. As a journalist and short story writer, Davis (1864-1916) was closely associated with the male chivalrous ideals most succinctly expressed in Theodore Roosevelt's doctrine of the "strenuous life." In his novel, Davis exploited the themes of strenuosity and virility to the fullest, explicitly linking them to American capitalist expansion abroad. One character in the novel explains that engineers "were fighting Nature at every step and carrying civilization with them. They were doing better work than soldiers, because soldiers destroy things, and these chaps [engineers] were creating, and making the way straight," but doing so without recognition. In fact, "the civil engineer," he continued, "is the chief civilizer of our century."[10]

The linkage between male conquest and the domestication of nature – at once sexual, female, primitive, and wild – found its most eloquent and unambiguous expression in Harold Bell Wright's best-selling novel *The Winning of Barbara Worth* (1911). The novel's title borrowed from the frontier manliness encoded in Roosevelt's *Winning of the West* (1896) and his other writings. In popular-culture representations, the West, in danger of becoming civilized and thus "feminized," figured as a place of male initiation by overcoming hardships." In Wright's novels, women's moral and civilizing influence threatens male enterprise and is therefore replaced by the kind of civilizing effort of engineers that Davis had in mind. Harold Bell Wright (1872-1944), a preacher turned writer of at least five best-sellers, reached millions of readers in American rural areas and small towns through a sophisticated advertising campaign of the Book Supply Company, a Chicago mail-order firm. Through his novels, readers learned to treasure middle-class values even if they had not yet joined the middle class in economic terms. Wright's writings romanticized the West, unspoiled nature, arduous labor, clean living, and neighborliness, all attributes of a Rooseveltian *Strenuous Life*.¹² His plots tried to reconcile engineering construction with the exploitation, scarring, and rape of the land, and the demands of capitalist individualism with the ideals of community and service. In these and other narratives, novelists portrayed engineers as battling greedy investors and taming nature, who fought both the forces of nature and the greed of financiers threatening their designs. Many of the plots turn on how the engineer-hero overcomes these barriers by virtue of his vision, hard work, and expertise. In the end, engineers of fiction reclaim nature and heroines, while they battle with irresponsible politicians and greedy capitalist adventurers.

The sales figures reflect the large audiences which the engineering novels were able to attract, but the authors' handling of the subject matter also accounts for their ongoing fascination with engineers in their plots. The outdoorsy masculine professional engineer carried great allure for male authors of popular fiction, who were writing and producing for a mass market, when popular magazines began to include articles on business, the professions, and politics in an attempt to court male readers.¹³ The introduction of new publishing techniques and the emergence of popular magazines transformed not only publishing but also the market position of writers in relation to audiences, editors, and reviewers. Before the Civil War, writing had been on the whole a leisure pursuit for which authors did not receive any payment; but by 1900, writers entered a well-established and centralized market facing pressures similar to those of engineers. Not surprisingly perhaps, in looking for a new male reading market magazines like *Scribners'* and *McClure's* were the first to serialize the engineering novels. Because the representation of the engineer could address male readers in a predominantly female reading public, as the father-son team understood so well, it had a specific market value to writers and publishers looking for new markets. Thus, engineers were not alone in their ongoing search for status and recognition.

KIPLING AND MARTHA'S MANLINESS

Best-selling authors like Harding Davis, Bell Wright, Fox, and Grey endowed engineers with a new cultural authority that elite engineers Van Deusen and Prior thought engineers so desperately lacked when the profession transformed into a mass occupation; but the English-American poet and writer Rudyard Kipling did even more. Of Thorstein Veblen's generation, Rudyard Kipling (1865-1936), the most popular poet and writer in the English-speaking world of his day, explored the male version of the engineering genre in a number of works. Like Veblen he was extraordinarily popular and a late Victorian helping to transform the age into one of Modernism. Kipling's readers were extraordinarily fond of quoting him, precisely because his work had different if contradictory layers and could be tailored to suit each occasion. Around the turn of the century, he reached stellar commercial success and could claim to be the most frequently quoted poet of the English-speaking world. He had to his credit 5 novels, 250 short stories, more than 800 pages of verse, and several books of nonfiction prose; 15 million copies of his collected stories were sold. His popularity paralleled the rise of modernism, to which his work bore little resemblance at first.[14] Most importantly perhaps for his popular success, Kipling presented himself as the voice of the ordinary worker and established the world of work as an appropriate subject for literature, in contrast to many contemporary writers who were primarily concerned with aesthetic subjects. He made labor the subject of his work and appropriated many working-class images of physical toil. Nevertheless, his celebration was thoroughly middle-class in appeal.

Kipling wrote many short stories and poems about engineers and their work such as *The Devil and the Deep Sea* (1895), "MacAndrew's Hymn" (1896), *007* (1897), *The Bridge Builders* (1898), and "The Sons of Martha" (1907) in the span of fifteen years. In these stories, he used technology in various ways: he inserted solid technical descriptions and anthropomorphized technical devices or employed engineers and their work in his narratives and verse.[15] The affection between Kipling and engineers was mutual. Not only did Kipling use engineers in his plots, but many engineers relished his work. He became their unofficial poet laureate. Engineers sponsored, quoted, appropriated, and reworked his verse when writing about themselves. Engineering magazines often published his poetry.[16] Ralph W. Lawton (1869-1943), an American civil engineer who managed the installation of sewer and water systems in India for the British government, wrote his autobiography in the form of a dialogue with Kipling's verse.[17] C.E. Moorhouse, a professor of electrical engineering, reportedly "made it a practice of commending Kipling [to his students] as a model of clarity in descriptive writing" because he considered Kipling's technical descriptions quite proficient.[18] In the pages of *Engineering*

Figure 25.   Kipling's poem, "The Sons of Martha" with focussed attention on the working-class male body illustrated in Gothic-symbolic style. *The New-York Tribune* (April 28, 1907).

News, Robert T. Gebler of the Technical Supply Company in Scranton, Pennsylvania, apologized to Kipling for attempting a verse very much like that of the master himself, which read in part: "But as I've often read it/The bloke who gets the credit/Is not the dusty khaki'd engineer/But the guys who 'ave the shillins."[19]

SUNDAY MAGAZINE
Of the NEW-YORK TRIBUNE

NEW YORK   APRIL 28, 1907

Containing RUDYARD KIPLING'S Latest Poem
THE SONS OF MARTHA

Figure 26.   Portrait of the most popular poet of the English-speaking world, R. Kipling, announcing his latest poem "The Sons of Martha," in *The New-York Tribune* (April 28, 1907).

These and other engineers helped rework the working-class body for middle-class consumption.

Engineers found in all of Kipling's work the recognition and the cultural authority they thought they lacked. But one poem in particular, entitled "The Sons of Martha", resonated deeply with them because of the class issues embedded in it, expressed in gendered terms. [Figure 25] In that poem he touched them in a visceral way. In it, he placed the issue of gender squarely in the middle of the discussion of work, and reinforced a male iconography that represented engineers as workers rather than as managers. Despite its ephemeral appearance one weekend in *The New-York Tribune, Philadelphia Press,* and the *London Evening Standard* in 1907, the poem acquired a subcultural following and became one of his more celebrated verses even though it never mentions engineers by name.[20] [Figure 26] Engineers saw themselves mirrored in the poem. It appeared to deal with engineers' social class and position, exalting "simple service simply given," as a middle-class ideal of work. Kipling's representation was in accord with nineteenth-century ideas of the work ethic ("It is their care that the wheels run truly; it is their care to embark and entrain/Tally, transport, and deliver duly the Sons of Mary by land and main."). The import of the poem lay in its celebration and validation of the thankless, subjugating nature of hard physical labor.

The wide circulation and recirculation of Kipling's poem show how his readers interpreted and rewrote it to suit the occasion.[21] "As in 'The Sons of Martha,' which my engineer-brother delights in," one reviewer reported, Kipling "has a way of pounding in his ideas with admirable economy of words."[22] In 1928, the contracting company of Mason and Hanger published a book marking the centennial anniversary of the firm; they called it *The Sons of Martha*, appropriating the image of labor and building, as engineers were bound to do. No explanation of the title was offered, since the writer apparently assumed that his readers would be familiar with the poem.[23] During the Great Depression the poem surfaced again, quoted in part in the editorial pages of *The New York Times* when a reporter attempted to capture the heroic efforts of the relief workers following the devastation caused by a hurricane that year: "It was then [at the moment of disorder] – that the Sons of Martha put on their boots and sou'westers and went out into the gathering darkness." In response to the editorial, one reader wrote ecstatically that the article was "one of the finest pieces of writing that has appeared in any paper. Every newspaper...should reprint it, and it should be read from every pulpit and from every radio station one Sunday in every year in honor of the men who do difficult things of life for no reason other than their belief in the necessity of doing them."[24] And as late as 1989, the Society of American Civil Engineers published an anthology entitled *Sons of Martha*, once again affirming their (engineering) readers' familiarity and ongoing identification with the poem.[25]

Engineers responded to the poem with pangs of recognition. In the poem, Kipling sang the praise of noble but unappreciated labor after the biblical story of Mary and Martha (Luke 10: 38-42). He turned the biblical Martha into a mother of men and fashioned a new icon of engineering masculinity: robust, strenuous, muscular, honorable, and anonymous. Kipling employed gendered images of work in establishing a contrast between producers and non-producers, Martha's and Mary's, men and women, and engineers and capitalists. In the circles of women readers, the story of Martha and Mary had served as a parable of women's socialization as fretting housewives. *The Woman's Home Companion*, for example, ran stories in which women who were called Martha invariably performed domestic duties. Even in a spoof on the image, Marion Harland's short story for the *Women's Home Companion* "Martha and Her American Kitchen" retained the essential outlines of the parable by associating it with women's socialization into good housewives.[26] Kipling had to dispell these notions.

In the traditional interpretation, Jesus's remark to Martha, that she should not be overly concerned with her domestic labor and should let Mary attend to her calling, became a canonical statement about the importance, if not the superiority, of spiritual labor for God's great work on earth. Contrary to this traditional exegesis, Kipling glorified menial work. He could only accommodate Martha by extricating her from the biblical role of feminine toil and recasting her in a modern, masculine role as an engineer. Instead of domestic representations of female toil, the poet laureate painted vistas of male work in engineering now associated with physical labor. In Kipling's reshaping of the parable, Martha had become a man. Thus, "The Sons of Martha" is based on women but is about men; it assumes ceaseless labor, but envisages toil as ennobling; it disparages the Word, but spiritualizes work.

Kipling added a class dimension to the portrayal of engineers; but to do so he mobilized images of women, however idealized, that engineers had so carefully ignored in their autobiographies. In choosing this parable, the poet used gender not to expound on women and their sons of flesh and blood, but rather to address issues of work and art. Kipling introduced gender into the discourse on labor whether he had intended it or not. As the title "The Sons of Martha" indicated, modern-day sons had inherited the tradition of toil from their biblical mothers.

The stanzas in which he extolled on labor rather than on the Bible were quoted most often. As Kipling realized only later, the explicit, irreverent treatment in his poem of those who did not get their hands dirty – God and Mary's sons alike ("They have cast their burden upon the Lord, and the Lord lays it on Martha's Sons,") – suggested to many readers that he had in mind the exploitation of workers by capitalists. In an introduction to a broadside reprint of the poem, Arthur M. Lewis explained why "it seems almost impossible to find this splendid poem in print nowadays." He had heard "that Mr. Kipling himself opposed its further circulation, supposedly on the ground of its class spirit." Lewis confessed he could not find any trace of class antagonism, but at least one reader-writer did.[27]

An anonymous poet in *The New-York Tribune* extricated the notion of labor from Kipling's ambiguous tangle and restored the class and gender hierarchy left dangling. In November 1919, the reader-writer rewrote Kipling's poem by reversing the roles of Martha and Mary, and entitled his "The Sons of Mary." The poet rejected Kipling's choice of Martha as an appropriate image of engineers-managers. In introducing the revised poem to readers, the editor explained that given "the labor conditions in all parts of the world," he believed the revision came "timed to the hour." The year 1919 was particularly dramatic for labor-management relations. In that year workers staged a series of strikes in response to skyrocketing increases in the cost of living during the war years. The many strikes such as the general strike in Seattle in February, the Boston police strike and the

nationwide steel walk-out, both in September, and the general coal strike in November left the middle classes in shock.[28] Reworking Kipling's poem to suit the
highly charged political climate of the postwar period, the anonymous writer defended management interests. He left no doubt about who he thought the real
workers of this world were and whose toil constituted true dignified labor. The
"Sons of Martha" lacked judgment, design, and initiative. Lest there be any doubt,
the author made it plain that he viewed workers as mere instruments with no technical knowledge, resourcefulness or motivation – a description absent from
Kipling's earlier evocation of labor. Instead, the poet in 1919 associated labor with
unions and a rioting rank-and-file, while manager-engineers embodied the mental
forces that directed and shaped production. Casting manager-engineers as the
Sons of Mary, the author rhymed:

> The Sons of Mary in all the ages have dared the venture and taken the chance;
> They explore earth's riches and plan the bridges, invent the machinery, design
> the plants.
> It is through them that on every work-day the Sons of Martha have work to do,
> It is through them that on every pay-day the Sons of Martha get every sou.

And while:

> They draft the maps and they paint the pictures; they carve the statue; the
> speech they speak
> ...the Sons of Martha seeking solely to do less labor for more per week.[29]

In short, in its 1919 revision, the poem became a vehicle for redefining class relations, shifting notions of labor from nineteenth-century ideas of male crafts – in
which at least a rhetorical harmony existed between mental and manual labor – to
twentieth-century managerial images of work. In 1907, Kipling's toilers still wore
badges of physical labor, but by 1919 this poet identified design as the most important marker of engineering identity. By his reversal of Martha and Mary, the
reader-writer cleansed Kipling's construction of work of any troubling working-class or female gender associations. Engineers were not alone in aspiring to a
distinction between craft and design, between workers and engineers, or between
manual and mental labor. This was also the year Veblen recast engineers as the true
producers of technical work.

Professional writers drew similar lines. If Kipling consciously positioned himself against the effete aestheticism of an Oscar Wilde by aligning himself with vigorous craftsmen, while keeping a safe distance from unionized workers, modernist
artists no longer saw themselves as craftsmen but as designers and professionals.[30]

They removed themselves from the crowds precariously associated with the mass of reading women. Modernists saw Kipling as representing everything they were not, but the gap between Kipling and the modernists was not so much a matter of content or style as a difference in the authors' relationships to their readers and the markets they sought to target. T.S. Eliot (1888-1965), the male modernist poet par excellence, once accused Rudyard Kipling of catering to the commercialized mass market, arguing that true artists would only write exclusively for the "one hypothetical Intelligent Man who does not exist."[31] Kipling did not simply function as a straw man for an emerging modernist agenda, but his greatest talent, Eliot accurately observed, was his ability to hold an audience beyond his own time. Kipling's very cordial relationship with a large and varied audience formed the basis for Eliot's critique and that of many male literary critics who came after him: commercial success and the mass of female readers became closely connected in the male modern mind.

WOMEN ENGINEER ALTERNATIVE PLOTS

If male writers sought a new professional identity on gendered terms in the belief that women dominated the literary markets, women of the world of letters had to deal with the social realities of earning a living in a man's world, even if they enjoyed stellar succes.

In the late Victorian era, women authors who explored the engineering genre problematized the newly forged alignment between engineers and male authors. The famed illustrator and writer, Mary Hallock Foote (1847-1938), the wife of the not-so-successful mining engineer Arthur De Winte Foote, was the first female author to venture into the male domain of engineering, and perhaps also the first to claim it as an appropriate literary topic for women writers. Mary Foote followed her husband in his mining career and entertained influential men of mining and geology including Clarence King, Samuel Emmons, Thomas Donaldson, and Rossiter Raymond. She used the engineering camps of Almaden, Leadville, Morelia, and Boise in the Western territories as the setting and subject matter of many of her stories and illustrations. [Figure 27] Foote did this so successfully that she became the sole breadwinner for periods of time when her husband's engineering projects failed in the decades from the late 1870s until the first World War. Living in the West, she provided East Coast readers of *Harpers' Weekly*, *St. Nicholas*, and *Scribners' Monthly* (later *Century Magazine*) with images of the West complete with homes and families which differed radically from the howling wilderness and manly adventure that Frederick Remington, Teddy Roosevelt, and Harold Bell Wright constructed in their depictions. They also differed from the colorless grids

DRAWN BY MARY HALLOCK FOOTE.

THE ENGINEER'S MATE.

ENGRAVED BY M. HAIDER.

Figure 27. Engraving "The Engineer's Mate," illustration for an article promoting settlement and exploitation of the Nevada desert for *Century Magazine* 1895 by author, illustrator, and engineer's wife Mary Hallock Foote, expressing her ambivalence about moving west with her husband. Courtesy of University of Amsterdam Library, Amsterdam, The Netherlands.

male engineers laid over the Western Territories in their autobiographical narratives.

Mary Foote's images of the West struck a responsive chord with her engineer-readers, whom she lovingly called "The Sons of Martha."[32] These Sons of Martha, she wrote in a reworking of Kipling's first stanza, "seldom saw themselves in print in any respect not connected with the paycheck or the announcement that the work didn't need them or had shut down." She received their letters in the engineering camp of Boise Cañon, Idaho, where she and her family suffered from isolation and the disasters that beset the irrigation project on which her engineering husband worked for almost 20 years. She reminisced how, "Most of [the letters] were in men's handwritings with queer postmarks, forwarded by the Century Company, from places as out of the world as the cañon itself. They followed every serial or short story dealing with the lives of our engineers in the field, and they came from mines and railroad camps on the far-flung lines of work, pushing new enterprises from Honduras to Manitoba."[33] But the engineers were critical readers who closely followed every detail of her ventures into the male domain: "They took these stories with delightful seriousness, not bothering about my technique but jealous for their own. They watched every term I used, every allusion where a pretender might slip up, when I undertook to speak the language of the sacred profession." In the engineering camps, her stories would be collectively read and discussed for technical content: "These letters would be signed sometimes by a group of names from the 'Old Man' to the 'Kid.' The Old Man, they said, had just been reading aloud to them the last story (or installment of a serial) under discussion, there being only one copy of the *Century* in camp; and would I please tell them how I came to know these things which the eye of woman hath not seen," she later recalled. "I answered delightedly and told them that I had married one of their lot and knew *them*, in their remotest hiding places."

As the wife of an engineer and as an artist who took herself and her writing seriously, Foote had wedded herself to engineering in more than one sense. Reflecting on her own difficulties and those of her family, she later playfully wrote, "often I thought of one of their phrases, 'the angle of repose,' which was too good to waste on rockslides or heaps of sand. Each one of us in the cañon was slipping and crawling and grinding along seeking to what to us was that angle, but we were not any of us ready for repose." This passage gave author and the historian of geology, Wallace Stegner, the title for his Pulitzer Prize-winning novel *Angle of Repose* in which the protagonist Susan Burling Ward closely resembles the life and correspondence of Mary Hallock Foote.[34] In many narratives – and Foote was not alone in employing the theme – the female protagonists routinely voiced criticism of industrialization and of the technical developement symbolized by engineers. Women writers had, of course, a body of literature and public discourse on which to base their criticism of the industrial commercialism associated with engineering. The early nineteenth-century canon of domesticity formulated values of female disinterestedness, service, and sacrifice to counterbalance and temper the male world of business and politics. But the emphasis in writing of these late nineteenth- and early twentieth-century women, who depended for their livelihood on publishing, shifted to a critical exploration of a male professional ideal. In the development of their plots, the authors usually harmonized the apparent conflict between ruthless male enterprise and female love, sacrifice, and civilization – all ideological attributes of the female culture of domesticity. And Foote seemed no exception, even if she had conquered engineering as her literary subject matter in a novel way.

In a central passage of Foote's short story "In Exile" (1894), written at a time when her husband was struggling with career setbacks and alcoholism, Foote pitted a female teacher against the engineer, the world of domesticity against the world of industrialization and immigration. Arnold, who is responsible for the construction of the mining camp's water supply system, orders the destruction of a natural spring, the site of their first budding romance: "The discordant voices of a gang of Chinamen profaned the stillness which had framed Miss Frances' girlish laughter; the blasting of the rock had loosened, to their fall, the clustering trees above, and the brook below was a mass of trampled mud. The engineer's visits to the spring gave him no pleasure, in those days. He felt that he was the inevitable instrument of its desecration."[35] The pervading discomfort in this passage – the presence of immigrant labor and the destruction of the spring – expresses ambivalence about the engineer's work, but, wedded as she was to the premises of engineering, Foote would never fundamentally question it, as the story's ending makes clear. Despite their differences, the engineer and the schoolteacher reconcile their worlds.

Foote depended on the values and the patronage of her East Coast publishers and readers and defended the expansion and the industrial development of the American West. As she was married to a mining engineer, with whose career her life was inextricably linked, it would have been impossible for her to draw any other conclusion. By the time she wrote "In Exile," Foote had decided to stick with her husband, even though she had seriously contemplated leaving him and his engineering schemes.[36] Where engineers described the same sites only with technical detail and without people, women writers like Foote filled the engineering camps they described with families and workers.

Another woman writer, the Westerner Charlotte Vaile (1854-1902), portrays one of the women characters in a similarly critical manner in her novella *The M.M.C.: A Story of the Great Rockies* (1898), in which the wife of a Colorado silver mine superintendent exclaims in the critical tone assigned to women, "The gold and silver might stay in the ground for all of me! I don't believe 'twas ever meant that men should spend their lives, burrowing like moles in the dark, for the sake of digging them out." Yet Vaile never allows her character to challenge fundamentally the inherently exploitative nature of the husband's work.[37] An even more critical distance from engineers' work and their emotional welfare is evident in short stories by two other women writers, Elizabeth Foote and L. Frances, published in 1905 and 1911 respectively.[38] With empathy, Frances focused on the discrepancy between a man's strenuous work and his poor emotional health in a story entitled, "The Engineer." And written from the "Girl of the Engineers" point of view, the highly educated New York reformer, librarian, and author Elizabeth Foote (b. 1866) – not to be confused with the famed writer and illustrator discussed above – exploited the same theme to the fullest and built a sarcastic story around the engineers' lack of emotional expressions and the resulting failure to communicate between the sexes. The emotionally segregated middle-class world led to close emotional relationships between women, binding them together in physical and emotional intimacy: middle-class women built a sororial world of love and ritual as historian Carroll Smith-Rosenberg demonstrated many years ago. To support her argument, Rosenberg culled some of her most salient quotes from the voluminous correspondence and dairies of engineer's wife and renowned author Mary Hallock Foote to her friend Helena spanning half a century from adolescence into widowhood.[39] Foote's Victorian life became emblematic for a genteel alternative female but segregated world as seen from the residential side of the engineering camps.

BURNING PROFESSIONAL BRIDGES

A younger generation of women writers including Anna Chapin Ray, Willa Cather, and Mary Pickthall went further than Hallock Foote in challenging the opposition between the male world of engineering and the female world of marriage, community, and art by their critical employment of a bridge collapse in their plots. They directly confronted the professional bridges that male writers and engineers constructing between the two professional cultures by burning them. All three women-authors suggested that the work ethic of men and the professional chivalrous code idealized in engineering were incompatible with genteel female culture. In the construction of their plots around disaster and collapse, Ray, Cather, and Pickthall had two other precedents, one historical and one literary. The widely publicized collapse of the Quebec Bridge in 1907 provided the dramatic backdrop. This cantilever structure spanning the St. Lawrence River was first touted as the greatest engineering achievement to date. When the bridge collapsed on August 29, 1907, killing 81 workmen and leaving an enormous tangle of twisted and broken steelwork, both the engineering press and the daily newspapers questioned whether the design for the enormous span had not gone beyond what was theoretically possible.[Figure 28 a and b] A well-publicized investigation eventually exonerated engineering theory and blamed the collapse on human error.[40]

In their novels, author of children's literature Anna Chapin Ray (1865-1945), novelist Willa Cather (1873-1947), and the poet and magazine writer Mary Pickthall (1883-1922) expanded on the conclusion of official reports faulting not just human but male error. Ray and Cather dramatically had their engineer-heroes fall with the bridges they had designed, while Pickthall made the collapse of the bridge the key metaphor in an engineer's failure to own up to his responsibility for the construction disaster.[41] All three women drew a close parallel between a faulty design in bridge construction and the flawed character of the hero. All three em-

Figure 28.   The building (a) of the Quebec Bridge spanning the St. Lawrence River. Its collapse (b) on August 29, 1907, came to symbolize faulty design of male character and professional ideals in several plots engineered by women novelists. Photographs both courtesy of Division of Engineering and Industry, National Museum of American History, Smithsonian Institution, Washington, DC. (neg. 91-6990 and 91-6981).

ployed their metaphor as a criticism of engineering and the process of industrialization embodied by engineers. And through their plot design, Ray and Cather also implicitly rejected the engineering professional as an appropriate professional model for writing women.

The literary framework of collapse as a theme was provided by Kipling's short story "The Bridge Builders" (1893). If "The Sons of Martha" resonated with engineers above all because of its celebration of producers and its condemnation of non-producers, Kipling's "The Bridge Builders" held a special appeal for these female authors, who reworked his theme in a differently gendered fashion. No doubt, as a literary theme the bridge offered many symbolic possibilities.

Set in India, "The Bridge Builders" first appeared in the *Illustrated London News* and was later published in *The Day's Work* (1898). The story's opening would later become a model for American writers, but the plot development also embodied preoccupations specific to British rule and economics in colonial India. Kipling begins the story as British engineers are finishing the construction of the Kashi railroad truss bridge over the river Ganges. The "bridge builders," Chief Engineer Findlayson, his assistant, Hitchcock, and Peroo, the faithful Indian assistant, are suddenly faced with a flood of biblical proportions that threatens the new bridge – a triumph of Western engineering ingenuity. The issue is whether Mother Gunga, the river, will accept the confinement of her floods and the marring of her banks by the new bridge. Although the depiction of the river Ganges as female (mother) was consistent with Indian cosmology, goddesses and animal gods were particularly disconcerting to the British colonial mind. Kipling exploited the association by extending this femininity to the colonial peoples as well, a trope that would become part of the cultural repertoire of Western racism.

Kipling specialized in abrupt and unexpected plot twists. Just as the bridge is about to collapse, the narrative shifts from engineering realism to the world of fables inhabited by Indian gods, demi-gods, and heroes and – this is uncharacteristic of the American elaborations of the theme – through the engineering crew's transforming experience with opium. Faced with the kind of crisis that would ordinarily invite the engineers of American fiction to flaunt their manly resolve, Kipling's Western bridge builders resort to drugs and are subsequently incapacitated. "It seemed that the island was full of beast and men talking," is all the engineer later recalls of the crisis, while the Indian assistant gains insight from the experience. Hearing Mother Gunga's case for the preservation of Indian tradition and religion, the council of Indian gods votes to override her by accepting engineering progress. The decisive argument presented in favor of the engineers is that the "fire carriages" (Kipling's mythological concoction for trains) will bring more pilgrims to the gods' shrines.[42] The final scene shifts back to a rather unfavorable depiction of the viceroy, the archbishop, and the colonial administration, who praise the engineers for their work, because thanks to the gods, but not to the engineers, the flood has receded and the bridge has been saved. The dramatic shift in narrative type from realism to fabulism and in character focus from the British engineers to the Indian assistant suggests the dissonance between the Western world of engineers and Indian society.[43] In other words, even though Kipling's story might be interpreted on one level as a simple endorsement of British colonial rule and industrialization – as many have argued – his plot development and resolution actually undermine or at least complicate such a conclusion. The plot subverted the general scheme of the story.

Because of their very multiplicity, Kipling's stories, like his poem "The Sons of Martha," offered writers a palette of possible themes and made his work eminently quotable. Anna Chapin Ray entitled her novel *Bridge Builders* (1909) but set it in Quebec, the site of the bridge collapse over the St. Lawrence River the previous year. In her novel, Ray reworked Kipling's gendered imagery. She transformed Kipling's conflict between traditional Indian society and Western technology by staging a dramatic contrast between the male profession of engineering and the female profession of writing. Educated at Smith College (class of 1885), Anna Chapin Ray wrote at least 40 volumes, some of them under the male pseudonym of Sidney Howard. She had spent most of her life in New Haven and Quebec, where she closely followed the reports on the collapse of the Quebec bridge. She was well prepared to write about the world of engineers: she regularly corresponded with her brother, Nathaniel Chapin Ray (1858-1917), a civil engineer working on railroad construction for the Burlington & Quincy, the Union Pacific, and the Oregon Short Line, in various parts of Iowa, in Boulder, Butte, and San Francisco. Over the years, brother and sister maintained a lively correspondence and exchanged details on their professional lives as a writer and a engineer.[44] She sent him clippings from technical journals. He advised her about marketing her work.

When Ray turned fifty, she wrote her own version of Kipling's *Bridge Builders*. Her plot revolves around two men, an artist named Kay Dorrance and an engineer named Asquith, both competing for the love of an exuberant young woman named Jessica. Her father, Peter West, a railroad contractor, appraises the manliness of his daughter's artist-suitor, observing to his surprise, "That fellow's very much a man, even if he does write books." He considers writing a "womanish" profession, contrasting it with Asquith's chosen career of engineering: "that's a man, all over, takes the best of a man's body and mind and soul." And Dorrance, the writer, muses, while assessing his rival in love, "What a man [the engineer] the fellow looked, dashing off like that. And after all, his was a man's profession, infinitely bigger, infinitely more virile than the mere knack of sitting in a corner and writing on a pad of paper." After an initial romantic rapprochement with the engineer, Ray's heroine falls in love with artist Dorrance, "a most unheroic for a hero," a man endowed with red hair, freckles, piercing brown eyes, brains, and money, but encumbered nevertheless by a "distressful limp," while the engineer is "tall, graceful, vigorous, virile."[45] The story's climax comes at the moment when the artist and the heroine are about to confess their love to each other on the very banks of the river the engineer has attempted to span. At this point the bridge collapses, and Asquith, the engineer and the embodiment of manliness and virility, disappears, crashing into the water with his bridge, which he has finally realized is his real love.

Ray apparently did not dare to dramatize the story's implications as Willa Cather did two years later in her novel *Alexander's Bridge.* For in Ray's novel, Jessica, portrayed as a New Woman, is a competent swimmer who scoops the engineer out of the water. Despite the rescue, she rejects the engineer's proposal of marriage, for she has come to realize that "this bridge of yours [is] the love of your whole life." Her father concludes likewise that "his very heart was in that bridge...He'll never care for any woman, as he cared for that steel arch." In the end, Jessica reconciles herself with the world symbolized by the genteel world of the writer. Thus, where Kipling's "Bridge Builders" ultimately, if precariously, accepted industrialization, Ray reworked her plot in order to reject unequivocally the male professional model of engineering. In particular, she rejected the engineer's total absorption in his work and his "marriage" to the company. As the artist concludes in the closing passages, "to me [writing] is the one great profession in the universe....That doesn't make me forget that the universe holds a few other things, though; love, family, friends."[46]

Willa Cather followed Ray's path to its logical literary conclusion. *Alexander's Bridge*, Willa Cather's first novel, went a step further in the rejection of the engineer as a professional model. In her plot, the collapse of the bridge kills the engineer. Familiar with *The Bridge Builders*, as a young writer Cather had admired Kipling calling him "a force to be reckoned with." Even if modernist male writers like Eliot rejected Kipling's mingling with the masses, Cather thought that "no man has ever written more persistently or more vividly of the affairs which engage the daily life of men." As a young writer, Cather had associated Kipling with Alexander the Great, the name she chose to give to the protagonist of her first novel. Unable to integrate the self of his youth with that of adulthood, Alexander, the engineer, dies with the bridge he has helped to design. The novel's premise – the engineer's faulty design and the collapse of the bridge – represent Cather's rejection, or so critics have argued, of the male perspective of Henry James and Rudyard Kipling, whom she had once considered her literary heroes.[47]

Cather's choice of an engineering theme for her first novel was a calculated one and the year 1912 was crucial. After years of working as managing editor of

140   *Making Technology Masculine*

McClure's from 1905 to 1911 – the magazine that had featured Hallock Foote and sought to appeal to male readers by including more articles on business, professions, and politics – Cather was well aware of the new trends in publishing. She left her editing job at the publishing company to devote herself full-time to writing as her profession. Considered in the light of Cather's literary career after her first novel, the theme of a collapsed bridge symbolized a break with her male mentors and cleared the literary path. Unencumbered by male precedents, she shortly thereafter entered a new domain she could claim as her own and found her own literary voice the next year in *O, Pioneers* (1913) and later in *My Antonia* (1919), staking out and reaching her own audience.[48] Instead of the Jamesian drawing room or Kiplingesque engine room, she recreated the lives of immigrants in Nebraska, where she had spent her youth. Deliberately chosing the engineer as the protagonist for her novel, she chose to display her competence in a male genre for the purpose of casting it aside, thereby signalling a new relationship with their readers.

In various degrees and with different points of emphasis, these engineering short stories and novels deal with the demands of total commitment to the company – a relationship the writers were wont to describe in terms of marriage and love. In the male authors' plots this commitment excluded women but in exploring the hero-engineer, women authors offered an alternative. They subverted the popular male genre of the period and rejected the engineer as a professional role model for female authors. Their engineer-protagonists failed because of shoddy workmanship, pressure of the market, or flawed characters that failed to integrate past and present, love and work. Chapin Ray, Cather, and others rejected not merely the industrialization engineers stood for, but also the masculine and muscular claims of the engineering occupation as an inappropriate professional mirror for authors.

As illustrated by the negative appreciation shown by male professionals for Cather's subsequent work, modernist taste makers rejected this female literary heritage and made male realism increasingly the standard for an American literary canon during the 1930s.[49] Despite Cather's phenomenal commercial success and critical acclaim from the literary establishment in the 1920s, she received increasingly disparaging reviews during the 1930s from a group of academic reviewers that was defining and establishing a distinctly male American literary canon. In response, Cather tried to forge her own relationship with her audience by circumventing the new male group of professional reviewers altogether, as her biographer Sharon O'Brien has argued. Despite, or because of, the large presence of women in the field, the process of professionalization of literary culture reinscribed it as a male province.

MODERNIST MOMENT: MACHINES, SEX, AND WAR

If popular writers of romance had staged engineers as a cultural hero, the modern art community did much to aestheticize the visual language of machinery and inscribe machines as explicitly male symbols. In celebrating the engineer, they followed the cue of popular romance writers. By the end of the first World War, engineers and the machines of industrial capitalism became closely linked in the literary imagination and visual grammar. After the economic crisis and cultural reorientation of the 1890s, the association between white Western men and machines was put center stage, finding its most powerful expression in the identification of male modern artists and writers with male-engineered machines. The modernists carried their rebellion forward in highly gendered terms in tone, image, and practice. A generation of modernist artists and writers who came of age just before the first World War began to caricature the sharply drawn Victorian divisions between male iconography of the technical and nontechnical world. They employed the machine simultaneously as a metaphor, model, and microcosm of modernity in the making.[50] In the years leading up to the war, many artists explored this discourse as a self-conscious way to becoming modern. By the 1920s, it had become quite common, if somewhat clichéd, to speak metaphorically about the *Machine Age*. This rhetorical position became a pillar of the modern understanding of technology and would be incorporated in the corporate image at the World Fairs in Chicago in 1933 and New York in 1939, no longer under the supervision of the Smithsonian but the National Research Council.

As abstracted and metaphysical entities, machines functioned in several ways in the visual language of modern art. Tools, devices, and machines like the drill, incandescent lamp, camera, and radio were generalized metaphors and lost the specific and local industrial surroundings of their production, yet acquired an aura of universal authority. With few exceptions, modernist artists represented technological devices from the consumer's point of view, with little concern for their production.[51] Such images neither smelled nor left behind any noise. This stainless consumer image of industrial production became incorporated in and mobilized for the new twentieth-century visual understanding of technology. But the pristine and sanitized visualization also jettisoned the graphic language of the Progressives' scathing critique of living conditions and omitted fingerprints left behind by industrial workers. Instead of heeding the tradition of the women's reform movement that was helping to built urban infrastructures, their formal language coalesced into a corporate engineering vision of processes and design. The visual language stressed mechanically and structurally infused masculine codings of prowess.[52] Dadaists neither invented these gendered and engineering images of machines, nor stood alone among modernist artists in choosing this subject matter. They merely, albeit brilliantly, exaggerated its latent meanings with biting visual irony. With anarchistic flamboyance and playful provocation, they further elaborated on the male alignment between engineers and artists that emerged in the popular literature during the previous decades as a strategy to escape from Victorian overstuffed female parlors. And male engineers, always seeking recognition, eagerly welcomed such a cultural stamp of approval.

As the U.S. began to prepare for the war, modern visual artists and writers were among the first cultural commentators to explore graphic and gendered language of what we now understand to be technology by linking men's control over machines and women. And when they did, they exploited its potential to the fullest. The Dadaist artists Francis Picabia (1879-1953), Marcel Duchamp (1887-1968), and Paul Haviland, all residing in the scaffolded city of New York during the 1910s, sexualized the machine metaphor by playing up gendered associations of machinery and appropriating the male iconography of engineers and mechanics. Like many of their modernist colleagues, they experimented with the tools and visual language of the engineering curriculum. Visual artists elaborated on the fascination of the male popular writers with the engineer as a male professional model, but like the mechanical engineer John Fritz and Kipling, Dada artists flirted with male blue-collar work by their sartorial identification with overalls. [53] With biting irony and exaggeration, the New York group of Francis Picabia, Marcel Duchamp, Paul Haviland, and others often used the sexualized machine metaphors as a means to bend Victorian notions of gender in search for modern models. As an act of transgression some modernist women like Frances Simpson Stevens boldly appropriated and exploited the new male subject matter of machines and engineering without any apology. After meeting the Futurists in 1913 in Florence, the young American Frances Simpson Stevens (b. 1895), who had been raised in the genteel halls of New England womanhood, explored the new machine language and trespassed into the male domain by her speedily painted 1914 work in oil and charcoal, *Dynamic Velocity.* [Figure 29] Her 1916 one-woman show in New York was a happening in good modernist fashion. Although reviewers did not know what to make of it, she received the most praise from a New York municipal engineer who recognized the engineering visual vocabulary in her work. [54]

By 1915, the year that generated many works of art in New York, the European war and the "sex war" were so thoroughly intertwined that one combat suggested the other, as Mina Loy's biographer Carolyn Burke has suggested. Sensing these vibrations in the air, Marcel Duchamp and Francis Picabia developed a formal vocabulary of what they called mechano-sexual metaphors, all showing a kind of sexual impasse and miscommunications between the sexes. Picabia's 1915 series of machine portraits of his artist friends in exile best represent the sexualized and gendered visualizations of the new emerging notion of technology in New York,

Figure 29.   Only surviving work of American futurist woman painter Frances Simpson Stevens appropriating the male encoded engineering style and subject for modernist women. Oil and charcoal on canvas Painting entitled "Dynamic Velocity; Inter-Borough Rapid Transit Power Station" of 1914. Permission Louise and Walter Arensberg Collection, Philadelphia Museum of Art, Philadelphia, PA.

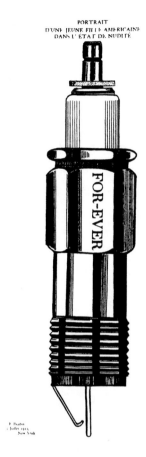

Figure 30.   Dadaist engineering portrait by Francis Picabia of "a Young American Woman in Naked Condition" drawn as a screwing device and spark plug representing the women colleagues of his modernist circle in New York. Published in Stieglitz's New York magazine *219* nos. 5-6 (July-August 1915).

the city that acquired a mythic status as the capital of modernity during the teens. Among them, Picabia drew photographer Alfred Stieglitz, Mexican caricaturist Darius De Zayas, and patron Paul Haviland. He laid out simple blueprints of ostensibly functional and operational devices like a camera, a radio diagram, and a lamp, which upon closer inspection turn out to be slightly out of joint and non-functional. Mischievously, Picabia called his drawing of a screwing device "*Portrait d'une jeune fille américaine dans l'état de nudité*" (or, "*a Young American Woman in Naked Condition*").[Figure 30] He played up the sexual transgressions and battles in their avant-garde community. In another portrait of his Mexican friend, entitled *De Zayas! De Zayas!*, Picabia drew a microscope with its gaze on a

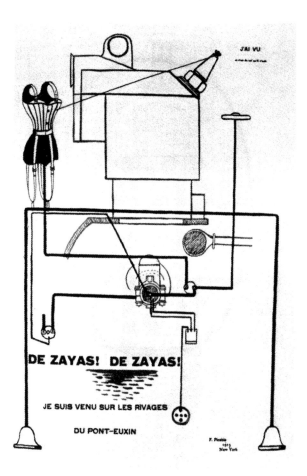

Figure 31.   Picabia's Dada portrait of his friend Mexican friend, de Zayas, engineers a tension between the male gaze and the female body in Dada visual vocabulary. Reproduced from Stieglitz's New York magazine *219* nos. 5-6 (July-August 1915).

corset. Here, Picabia juxtaposed a seriously encoded male instrument with a symbol of female frivolity and sexuality. He surely meant to parody the newly drawn gendered boundaries between the technical and the non-technical. His patent-like depiction of the corset also alluded to a long-standing tradition of women's inventing practice and turned it into a figure of irony."[Figure 31]

Marcel Duchamp, following Picabia's ironic and sexualized depictions of machines, engaging in one of his most elaborate and extended modernist jokes. He entitled his more notable art piece "*The Bride Stripped Bare by her Bachelors, Even,*" on which he worked for eight years (1915-1923). It represented the bachelors as mechanical devices going through a seemingly functional treadmill, while it portrayed the bride as a static, separate, and shapeless space. Whatever his intent – the

subject of heated debates in the literature – Duchamp engineered a tension between the enigmatic messages of gender relations and the simplicity of the lines
that mocked, modelled, and commented on the male engineering ethos of simple,
clean, functional, and minimalist lines.[56] To symbolize sexual intercourse in such
bare, stark, and mechanistic terms was a severe affront to Victorian aesthetics and
epitomized the modernist rebellion against it.

In the same year that Picabia drew his mechanical portraits and Duchamp began to engineer his mechanical *Bride*, the Dadaist Paul Haviland, lawyer, patron,
and member of the Alfred Stieglitz circle in New York City, elaborated on the new
and explicitly male vocabulary of technology his friends were developing. Mar-
veling at such new technologies as the hand-held camera, the phonograph, and
electricity, Haviland called for the mastery of men over machines. He offered an
explanation for men's technophilia: "the machine is his 'daughter born without a
mother,' that is why he loves her. He has made the machine superior to himself.
That is why he admires her." To emphasize the sense of male mastery implied in
his metaphor, he concluded: "she brings forth according to his conceptions."[57] He
likened machines to women's bodies – something already implicit in the Spinning
Jennies of the textile mills of an earlier age – but Haviland made an explicit
gendered connection of what would become a common trope: men's mastery over
machinery and women. Men were producers, while women were merely instruments.

If the Victorian male and female writers had worked separately and in isolation,
the modernist men and women mingled freely, often visiting each others' studios
and apartments on a daily basis. The French-speaking women artists including
Mina Loy, Picabia's wife Gabrielle Buffet, and Juliette Gleizes who frequented the
Arensberg soirees in New York all felt awkward, distressed, and ill at ease when
watching their male colleague Duchamp encourage their American female colleagues to enact their loss of sexual innocence and perform his extended sexual linguistic plays at the expense of the American women who were less versatile in
French.[58] To avant-garde women it might have been liberating, but to contemporaries, sexual liberation, women's suffrage, artistic innovation, and political protest were all part of the same controversial landscape. To be sure, Mina Loy,
Frances Simpson Stevens, and Baroness Elsa von Freytag-Loringhoven fought
their own battles to become modern as women and as artists, but increasingly recognition for women artists had come just as modernists began to question the
academy system. Modernist women also flouted bourgeois manners and criticized
the hegemony of classical antiquity, romantic love, painting from the model, and
established cultural institutions like art academies and museums, but their gen-
der-bender battles took other forms. Not in the least because for women the act of
trespassing had a different price tag attached than it did for men.

Frances Stevens's older friend, the English-American artist Mina Loy, a painter, poet, and playwright, explored but ultimately criticized this new modernist sexualized language. In her *Feminist Manifesto*, she wrote that she believed men and women were enemies: "The only point at which the interests of the sexes merge is the sexual embrace." And in a poem, entitled "Human Cylinders," she first adopted, then disregarded the Futurist vision which likened humans to machines in answer to her futurist lovers Martinetti and Papini. Intercourse was just a collision of bodies and "Love with me is a mechanical interaction," she wrote.[59] As Mina Loy understood only too well, gender bending carried a higher price for women than it did for men artists. Nurtured in distinct women's traditions, Victorian women writers had been able to engineeer their own plots, but for modern women artists whose work and social life were so closely linked to men, it was much harder.

In the end, modern artists too helped reinscribe rather than subvert the male iconography in technical objects through their graphic, often sexually and gendered explicit language and images. Despite the ironic sexual negotiations and the explorations by other women artists, by the 1920s the machine aesthetics of art was thoroughly revitalized as a modern male icon in which a new generation of public intellectuals, social scientists, scientists, and engineers began to participate with new vigor. In the end, gender bending too was not the same for all.[60]

5

Women Reweaving Borrowed Identities

I f women writers, artists, and activists articulated an alternative language, their slide-rule sisters within the engineering occupation emulated rather than questioned male models of professionalism. American women engineers have left few written traces of their existence, unlike their male colleagues or their sisters in the world of literature. They could have flaunted their pioneering struggles in the manner typical of autobiography, but they neither adopted autobiographies as a form of self-expression nor created alternative plots of their own. Steeped in the ethics of self-discipline, stoicism, and overqualification, they had few narrative devices available to them.[1]

Among American women engineers, only one started an autobiography, Nora Stanton Blatch (1883-1971), whose rich feminist heritage enabled her to envision a narrative device in which to frame her life story.[Figure 32] As a third-generation feminist fighting for suffrage, Nora Blatch could project herself into a well-defined feminist genealogy. She descended from a line of famous feminists – her mother, Harriot S. Blatch, and her grandmother, Elizabeth Cady Stanton – and campaigned for suffrage at Cornell University, where she had chosen civil engineering as her major because, she said, it was the most male-dominated field she could find. Her generation of women engineers grew up in the nineteenth century, when the bond of solidarity among women was more firmly entrenched, but Blatch went a step further than her contemporaries. She contested the American Society of Civil Engineers (ASCE) on grounds of gender discrimination when in 1916 it tried to bar her from full membership; moreover, she campaigned for pay equity between men and women through the National Woman's party for many years. Where Blatch found literary and organizational models in the feminist movements of her mother and grandmother, she found none among her women colleagues in engineering.[2] Raised on the expectation of the government's propaganda for more technical personnel during the first World War, the next generation of women engineers grew up in an era where suffrage had been won and professional women seemed to be making headway. They adhered to a belief in gender neutrality and the merit of professionalism. Few women engineers publicly rallied to feminist causes.[3] This disinterest was mutual. The American women's movement invested

Figure 32.   Nora Stanton Blatch, civil engineer and descendant of two generations of women rights activists shown on suffragist campaign on horseback in New York State in 1913, challenged the engineering establishment in 1916 on charges of discrimination. Reproduced from *Civil Engineering* (1971). Courtesy of Delft University of Technology, Delft, The Netherlands.

in the sciences rather than in engineering because it lacked the same cultural authority. American women's education activists and female philanthropists neither paid any special attention to the engineering profession as a vehicle toward women's equality nor helped establish separate engineering institutions.[4] Rapidly becoming a mass occupation, engineering failed to attract bright American young women of high standing looking for a suitable vocation. On average, women in engineering came from a higher class background than their male counterparts.

Most American women engineers ignored the kind of bridges Nora Blatch tried to build between the women in the technical field and those working in the woman's movement. They honored the model Nora Blatch's contemporary Lillian Gilbreth (1878-1972) offered instead. Gilbreth borrowed Rudyard Kipling's rewording of the biblical allegory of Martha ("simple service simply given") that was supposed to be the inspiration for women engineers. Gilbreth, earning a

Ph.D. in psychology, had received her technical knowledge and her legitimacy in engineering through her husband – a "borrowed identity" she expertly managed. Although actually a widow for the larger part of her life, she projected herself as a married career woman.[5] Avoiding the open confrontation for which Blatch opted, she advocated a professional strategy for women engineers based on hard work, self-reliance, and stoicism. Her employment of Kipling's poetry embodied her conservative strategy and showed all the ambivalence of the position of women engineers as rank-and-file members.

Such strategy involved tactics of "quiet but deliberate over-qualification, personal modesty, strong self-discipline, and infinite stoicism," as historian of science Margaret Rossiter has pointed out to describe "the classic tactics of assimilation required of those seeking acceptance in a hostile and competitive atmosphere, the kind of atmosphere women heading for bastions of men's work encountered at every turn."[6] Indeed American women engineers maintained their loyalty to male models of the profession at great personal cost. Vera Jones MacKay, a chemical engineer who had managed to find work on pilot plants for fertilizer production with the Tennessee Valley Authority, recalled the painful memories when looking back on her career in 1975: "it is hard to discuss my working days as an engineer without sounding like one of the most militant of the women's libbers."[7] Jones's public admission of the personal costs involved in her career choice is unusual because most women engineers kept a stiff upper lip. Yet her sentiment offers a rare glimpse into the struggles and strategies of women engineers in the period before affirmative action. Most women engineers who preferred as their role model Lillian M. Gilbreth over her contemporary Nora Blatch followed this strategy. They drew their literary models and organizational forms from their engineering fathers rather than from their feminist sisters. They cultivated silence as a survival strategy and ventriloquized, but never directly articulated their discontent. As rank-and-file members of the profession working for corporate and military establishments, women engineers became invisible not only to themselves, but also to history.

SURROGATE SONS AND THE INSIDE JOB

A great many American women, most of whom never appeared in any statistics, claimed family ties to engineering through their fathers, brothers, and husbands. They found their pathway into engineering through what might be called patrimonial patronage and matrimonial sponsorship. Supported by kinship ties, such familial patronage and matrimonial sponsorship often offered relatively easy access but also resulted in what German historian Margot Fuchs has called a borrowed identity. Most did not have female models but looked to their fathers and brothers for orientation. Their identity as engineers was therefore largely one "on loan," even if some, like Gilbreth, managed to stretch the terms to socially approved limits.[8]

Formal training might have been an important credential for continental Europe and served as a wedge into other professions in the U.S., but it did not play such a decisive role in employment opportunities in American engineering before 1945. By the end of the Forties and beginning of the Fifties, only 55 percent of American men – and even fewer women (20 percent) – in engineering had completed an engineering education. In the nineteenth century but probably also well into the twentieth century, a few hundred women continued to manage their husbands' engineering work after his death, having received enough informal technical training to call themselves engineers.[9] They acquired technical knowledge on the job or through an informal system of education within family firms without ever attending a specialized school. Famous examples of women who learned the trade through their family or husbands include Lillian Gilbreth and Emily Warren Roebling (1841-1902), who kept the family firm going when illness incapacitated her husband, a chief engineer, from leaving his house. She acted as his proxy throughout most of the building of New York's Brooklyn Bridge during the 1870s and 1880s. Trained in mathematics, Roebling learned to speak the language of engineers, made daily on-site inspections, dealt with contractors and materials' suppliers, handled the technical correspondence, and negotiated the political frictions that inevitably emerged in such a grand public project. The Brooklyn Bridge had been a Roebling ideal on which the family's fortunes depended, and Emily Roebling had been her husband's proxy for decades. Less famous than Roebling, most wives worked in anonymity in family businesses. As late as 1922, a woman active in civil engineering wrote the editor of *The Professional Engineer* that she greatly appreciated that the journal finally acknowledged the wives of engineers without specialized degrees: "My training in engineering began with marriage and I have filled about every job... from rodding and driving stakes to running a level party, or setting grade and figuring yardage in the office."[10]

These women would be largely forgotten. Lillian Gilbreth, however, became America's most celebrated woman engineer in part because she managed to expand the limits of her borrowed identity into socially permissible terms. Frank Gilbreth's untimely death in 1924 might have been devastating for a mother with twelve children but, aided by a team of domestic hands, it also allowed Lillian Gilbreth to enjoy considerable freedom in her role as a widow for nearly fifty years. She expertly managed her image, fostering publicity that cast her in the role of a married career woman. This public persona provided perfect protection against the possible disapproval of her career ambitions. Similarly, she allowed her marriage, long after Frank Gilbreth's death, to be promoted as the most efficient way of life: as a business-love partnership – analogous to her husband's advocacy of efficiency by performing tasks in the "one-best-way."[11] As Gilbreth's strategy shows, women might also have invoked family connections as a way of protecting them from public scrutiny. Newspaper reports and government propaganda keenly played up women engineers' strong family ties to men in engineering as a way to ward off the possible threat of these female incursions into the male domain. The effort to 'domesticate' women's talents into familiar categories prevailed during the second World War, when war propaganda emphasized women's family ties to engineering. Most of the available biographical information on the social background of women engineers was generated as part of the war propaganda campaign during the Forties; historical narrative sources on women engineers therefore tend to overexpose women with family connections. Nevertheless, it is evident that formal engineering education with or without a degree in hand could be particularly useful for some, like the daughters of proprietors of small manufacturing firms. Beatrice Hicks trained at the Stevens Institute of Technology to become first chief engineer, then Vice-President, and finally owner of her father's Newark Controls after his death. After their graduation in engineering, Jean Horning Marburg supervised the plant construction for her family's mining property in Alaska, while Florence Kimball worked at her family's elevator firm, drafted plans for the remodeling and building of its real estate property, and drew several blueprints for patents – the most exacting of all draftmanship. Small family firms like the Kimball and Horning companies not only tried to maximize production and profits. They were also in the business of perpetuating a family legacy.[12]

Succession in patriarchally organized family firms was exclusively an affair between fathers and sons, but circumstances sometimes pushed daughters into the positions of surrogate sons. The most celebrated and best-documented case is that of Kate Gleason (1865-1933), the eldest daughter of William Gleason, who had started his own toolmaking shop, the Gleason Gear Planer Company in Rochester, New York, that would be one of the largest of its kind. Gleason combined her feminist independence with acute business sense and family interests which appealed to both women engineers and popular writers. Encouraged by feminist Susan B. Anthony's example and prompted by the early death of her half-brother at the age of 20, Kate Gleason began to work for her father after taking courses in mechanical engineering at Cornell University and the Mechanics Institute in Rochester. Her training followed the course of many sons of other family manufacturing firms, who were no longer expected to master a craft completely, but to have a working knowledge of all the various aspects of the firm. Gleason was her family firm's business manager for many years when the business expanded dramatically and successfully into a major player in the industry.[13] Patrimonial patronage thus encouraged daughters like Kate Gleason to seek formal education with or without completing a degree because it fit into a family business's strategy.

For similar reasons husbands encouraged their wives to seek formal training. Such active matrimonial sponsorship not only provided women with a legitimacy their engineering accomplishments would otherwise have lacked, but also offered them the hope of establishing a family firm in partnership with their husbands. The pooling of resources of man and wife in an enterprise offered the opportunity for a partnership of business and love. Sometimes, however, engineering marriages offered an advantage that could turn into a liability. Many women met their partners at college or in the field of engineering, allowing them to enter into male social and study circles otherwise closed to them. But because of the inherent power relationship, the built-in mentorship in such relationships could turn into a distinct disadvantage for the wife's career advancement later on when they questioned the terms of matrimonial sponsorship and their borrowed identities that went along with it.[14]

As a young feminist activist and engineering graduate, Nora Blatch and her husband, the engineer and inventor Lee De Forest, first shared in the excitement of new emerging technologies such as the radio, but in the end they disagreed about who was to shape and direct the possibilities of these novel developments. On their first meeting, Blatch "tremendously admired" the young radio inventor Lee De Forest and explained that "a life in the midst of invention appealed to me strongly."[15] For his part, De Forest thought "destiny" had brought her to his door and pursued her relentlessly. In desperate need of money for various ventures, he received funds from his future mother-in-law Harriot S. Blatch, while Nora's technical training, her love for music, and the connections with the New York powerful, brought enormous technical, financial and social resources to his flagging career. No doubt seeing an opportunity to fulfill her life's goal of combining career and marriage, Nora Blatch fell in love with Lee De Forest, took extra courses in electricity and mathematics with Michael Pupin, a well-known New York electrical engineer, and worked in De Forest's laboratory on the development of the radio. Together, Lee and Nora were able to air the first broadcasts of music and conversations in the New York area. On their honeymoon to Paris, the newlyweds seized the opportunity to promote their wireless phone by a demonstration from the Eiffel Tower, organized through Blatch's family connections.[16]

Both Blatch and De Forest shared an excitement about participating in the new technological developments with their contemporaries. To Harriot S. Blatch – and there is no reason to believe Nora disagreed with her mother on this issue – technologies such as the radio were new tools for women to use for their own ends. At one of the promotional experiments for the "wireless phones" in New York in 1909, Harriot, Nora, and Lee were positioned at one end of the transmitter at the

Terminal Building, while a group of women's students from Barnard, their physics professor, and some male interlopers from Columbia stood listening at the other end at the Metropolitan Life Building. "I stand for the achievements of the twentieth century," Harriot Blatch declared in the first message transmitted. "I believe in its scientific developments, in its political development. I will not refuse to use the tools which progress places at my command... not forgetting that highly developed method of registering my political opinions, the ballot box." Since the transmitter was only a one-way communication, she continued uninterrupted – despite a protest from a male student from Columbia: "That is a mean way to talk at a poor chap when he can't say anything." Believing that technical modernity was inextricably and inevitably linked with politically progressive ideas, she continued: "Travel by stagecoach is out of date. Kings are out of date: communication by canalboat is out of date; an aristocracy is out of date, none more so than a male aristocracy."[17] The speech was used by De Forest and his business agent to sell stock of his Radio-Telephone company to suffragists and their supporters.[18]

Even if Nora Blatch and Lee De Forest shared in the excitement of the new technologies, disagreements emerged over the financial status of the firm once they had married and their child was born. De Forest so strongly opposed to his wife's management views on the family firm and the work she continued to do in engineering that it caused their separation in 1911. Explaining his divorce, De Forest told reporters of a national newspaper that "his matrimonial catastrophe was due to the fact that his wife... had persisted in following her career as a hydraulic engineer and an agitator [for women's suffrage]... after the birth of her child."[19] He warned other men against employing their wives, conveniently omitting all mention of Blatch's technical and financial participation in his ventures. Eventually, Blatch started her own architectural firm with family capital. It allowed her to remain independent from partners like De Forest and from the corporate employers she had earlier learned to avoid. Although offered more than expert labor, De Forest and other husbands were interested in a joint venture but not an equitable partnership with their wives. To De Forest, who insisted that he wholeheartedly supported suffrage for women, admired Nora's intelligence, and enjoyed her technical training, his wife's greatest offense had been that after marriage and motherhood she had rejected an on-loan identity and continued to assert her feminist heritage of three generations and her own career as an engineer.

SCHOOL CULTURE AND THE STRATEGY OF OVER-QUALIFICATION

Family businesses were based on a form of engineering knowledge which linked them to the patriarchal authority of the traditional workplace where class shaped the relations between management and labor. Formal education, by contrast, was to be a more democratic form of knowledge accessible to all, but it was still in need of establishing and reproducing its own male model of authority. In the decade following the American Civil War, diversity and openness characterized American engineering education, but it nevertheless came to be bound by gender and race. Hailed as the landmark legislation that pushed higher education to unprecedented coeducational levels, the Morrill Land Grant Act of 1862 helped establish several schools of engineering at land-grant state universities, colleges, polytechnic institutes, and private universities throughout the land. Its drafters had intended it for the education of the children of farmers and industrial workers, but had not stipulated the character of "agricultural and mechanic arts." In the early days, women, workers, and farmers attended courses given at the land-grant colleges institutions like MIT. Industrialists had been the first to support education of workers and women viewing them as a potential disciplined work force.

This broad commitment of the Act changed in the course of the century. "The agricultural and mechanic arts" often came to mean industrial rather than agricultural education, technical rather than artisanal training, and school-based engineering rather than a British-style apprenticeship. Engineering educators began breaking with the traditions of vocational training, managing to seize all the attributes of scientific rhetoric. The push of upgrading the field through the infusion of professional ideals resulted in the masculinization of the higher education of engineering, sending women into separate fields of chemical lab work or home economics.

Before this closure, American women were welcomed as special tuition-paying students when engineering educators sought to increase their enrollment figures for their newly minted programs. American women had free access to primary and secondary education and came relatively well prepared compared with their sisters elsewhere. In particular the recently minted programs were more welcoming than the established institutions. Thus, the coeducational land-grant institutions and state universities showed a more favorable attitude towards women's higher education in engineering than privately owned and sex-segregated institutions like denominational colleges, military academies, and high-status private schools. The state-sponsored land-grant institutions (e.g. Purdue, MIT, Iowa State, Ohio State, Cornell, Berkeley, and the Universities of Washington, Illinois, Colorado, Michigan, and Kentucky) and many municipal universities (the Universities of Cincinnati, Louisville, New York, Houston, and Toledo) pioneered in coeducation in engineering. Even some mining schools admitted women to their engineering departments.[20]

In the pre-professional era, when neither engineering institutions and occupational clubs had yet raised their standards to meet the prestige of the other professions and home economics had not yet been established as a separate field for women interested in technical fields and applied sciences, pioneering women students began to graduate in engineering from the 1870s onwards. Even so, they received mixed messages. Engineering educators searching for higher enrollments might have admitted some women to their programs, but aspiring women faced outright discrimination at every turn, requiring a stamina not all women could muster. A complete set of data on the enrollment and graduation figures of three schools (Ohio State University, University of Alabama, and Stanford University) suggests – not surprisingly perhaps – that the dropout rate for women was 25 percent higher than for men, 50 instead of 40 percent of those enrolled.[21] Even women who managed to complete their course work did not always receive the official recognition they deserved. The experiences of Lena Haas at Columbia, Eva Hirdler at the University of Missouri (1911), and Mary and Sophie Hutson at Texas A&M (1903) are telling examples of women students who satisfied all their requirements without receiving the appropriate degrees during a period when engineering educators tried to raise academic standards to compete with their colleagues in the humanities.[22]

Facing discrimination, women engineers paired their strategy of overqualification with stoicism. The experienced mechanical engineer Margaret Ingels warned in the 1930s that a woman engineer "must in many cases work even harder than a man to build up confidence." Two decades later another woman found the situation unchanged and concluded that "a dedicated woman can succeed [but has to] run twice as hard as a man just to stay even."[23] Women who were willing to fit into the tightly knit male world of engineering could force the doors slightly more ajar by concentrating on their math abilities and doubling their efforts. Many women of the early generation opted for multiple degrees with which to scale the academic walls of engineering.

If women engineering students in the nineteenth and early twentieth centuries faced formidable difficulties, the lack of preparation in mathematics does not seem to have been one of them. In high school, for one, American girls and boys received an equal amount of instruction in calculus and geometry and came relatively well prepared.[24] Moreover, because women who entered engineering tended to come from higher social backgrounds than their male counterparts, they often came academically better prepared. A few decades ago, sociologist Sally Hacker argued that the high standards of math in engineering education effectively served to exclude women, but in the transitional period from a proto-professional to a professional era, math offered a window of opportunity for those women interested in a technical education, however briefly. Before the second World War, an understanding of mathematics was required for practicing engineering, but in America it never formed the kind of obstacle or rite of passage that it would later when it became part of the tactics to raise the standards of engineering, Hacker observed at schools like MIT during the 1970s. On the contrary, many women who went into engineering could claim superior ability and knowledge in mathematics. The increased importance that engineering educators placed on mathematics as a means of upgrading the profession might have been a major hurdle to many engineering students with average ability – women or men – but it also acted as an advantage for brilliant women in a school culture that stressed academic skills over hands-on experience. Exceptionally competent women like Elsie Eaves (1898-1983), Alice Goff (b. 1894), Dorothy Hanchett (1896-1948), and Edith Clarke (1883-1959) used their mathematical skills and multiple degrees as a wedge into engineering work and mobilized them as a shield against outright discrimination.[25] For brilliant women, excellence in math proved to be a window of opportunity at first.

Educational reformers like Thurston who sought to upgrade engineering training with a new emphasis on mathematics, history, and the humanities faced a dilemma. Their form of engineering knowledge was not linked to the patriarchal authority of the workplace where class shaped relations between management and labor, but based on the new cultural authority of science and math. Not only were academic engineers often accused of failing to prepare their students to face the reality of the production floor, but academic ideals threatened to become associated with gentility and femininity. In this balancing act, engineering educators became the most articulate purveyors of an academic male esthetics that stressed hands-on experience and a slap-on-the-back kind of manliness. Many engineering educators tried to imitate "the methods and manners of real shop-life" in college shops that housed steam engines, blacksmith tools, foundries and the like. Here hands-on experience could be acquired while preserving academic ideals. However well-equipped, the problem with the college shops was that true confrontation with the attitudes of independent workers and bullying foremen could neither be simulated nor tested. The ability to "handle men" remained the true hallmark of the successful engineer of a management professionalism of engineering. In the schools of engineering, this managerial ideal of engineering balanced precariously between working-class manliness and academic gentility. Within the walls of academia, work in the laboratory, the shop, and the field clad in overalls with a knack for tough jokes added luster to the male rites of passages into the profession. It served not only an educational goal but also sought to enhance the prestige of engineering education.

In these environments, women students were encouraged to take math classes but often excluded from taking shop or field trips to factories mandatory for graduation. Around 1904, when Nora Blatch's classmates prepared to pose for a photograph showing them working as civil engineers in the field, they arranged for a male friend to date their female colleague on the day of the photo session so that she would not be in the picture. They thus deliberately excluded her from this male rite of passage and erased her from the visual historical record. In 1925, MIT professors prohibited Olga Soroka from participating in a field trip required for graduation in civil engineering. Her professors organized a special internship with the New York subway for her instead, considering it more appropriate to contemporary (but always changing) definitions of women's public behavior. Anna Lay Turner, a chemical engineering student at Rice University in 1924, recalled that women were tolerated in genteel academic environments, but barred from mechanical labs. To put on overalls was to challenge prevailing codes. While engineering educators might encourage women students to take courses involving the sciences and mathematics, they tried to bar them from practice classes and laboratories, even when these were necessary for their graduation. All these prohibitions expressed the strong resistance to any female incursions into this specifically male engineering rite of passage of class.[26]

The workshop and building sites thus functioned as a way of screening out women "[for] it must be clearly understood," as one critic of women in engineering and other technical occupations wrote in 1908, that "the road to the drafting board and the laboratory of the engineer lies through the workshop, and workshop practice means hard work and blistered hands, not dilettante pottering and observation."[27] Women might be competent in drafting, calculation, research, and analysis, as employers testified in the 1920s, but sweat, dirt, and calluses made the engineer a real man. Or, in the words of one scholar, "if science wears a white lab coat, technology wears a hard hat and has slightly dirty fingernails."[28] Ideally, middle-class men belonged on the production floors and building sites where they managed other men, while women dealt with more technical details in genteel environments. But this was true only as an ideal to be aspired to.

FOOT SOLDIERS OF BUREAUCRACY

Most women engineers were employed by the emerging military-industrial complex. Women also found their way into engineering through what should be called corporate and federal apprenticeship, particularly when corporations and the government worried about a shortage of technical personnel in times of war and competition with foreign countries. Among entry-level jobs, women made the most headway in the laboratory-oriented and newer fields which did not carry gender codings at first and required more academic skills: chemical analysis, electrical, and – after the second World War – aeronautical engineering. If small businesses provided a way into engineering for women with family ties, the large emerging corporations did so for women without family capital or resources, to whom mechanical engineering – steeped in craft traditions – remained a closed shop. This is not to say that women could not be found working in mechanics shops: during the first and second World War, corporations hired working-class women as lathe and punch press operators and as assembly workers. Mechanical engineering implied a different class than the engineering trades, however, one that involved supervision.[29] These temporary encouragements women received in pursuing engineering were ambivalent to say the least.

For employees of large corporations without family connections or capital, an engineering job held the promise of promotion, even if it became more a vision than a reality in the course of the twentieth century. The formal and bureaucratic rules both institutionalized endemic gender discrimination and helped to secure better opportunities for women engineers without the proper family ties than the informal but patriarchally infused rituals of firms where shop-floor culture encouraged male patterns of advancement. These were open only to the few women like Kate Gleason who could crack the male code of the shop floor by invoking an authority stemming from family ownership. The growing importance of formal rules and the move toward professionalization in the twentieth century proved to be a two-edged sword for women engineers who chose to enter the profession without capital or connections. The two world wars – and the state – offered windows of opportunities, but not full-fledged careers. The war economy also institutionalized and created new discriminatory practices.

Women might find formal education a viable means of access to entry-level engineering jobs, but those who excelled academically did not fare well subsequently in their engineering careers, either because of matrimonial disloyalty or because of male codes of the workplace. Highly trained women including Elsie Eaves, Olive Dennis, Patricia Stockum, and Mabel Macferren, all of whom had earned two or three degrees and showed the stamina to succeed, found that this initial advantage turned into a liability once they entered the workplace. In the workplace environment male codes of managerial command and hands-on experience determined one's professional standing, not academic excellence. Many overqualified women ended up either as (high school) teachers in mathematics and sciences or as calculators in corporate offices and at research institutions. Dorothy Tilden Hanchett first trained in civil engineering at the University of Michigan ('17). No doubt she believed that additional M.A. and Ph.D. degrees at Columbia University ('27) and Logan College ('45) would help to advance her career. Instead, she ended up at Battle Creek High School as head of the math department. In the aftermath of first World War, Hanchett and many other highly qualified women found that government propaganda had contained more rhetoric than reality. They were forced to accept temporary teaching jobs in elementary and high schools, teaching instructorships at engineering colleges, or editing positions in professional organi-

zations.[30] Thus, the tactic of obtaining multiple degrees did not guarantee employment.

In times of economic bust, only government highway projects and bureaucracies could offer academically trained women employment, however ill-paid. Even
if proportionally more women trained in civil engineering than in any other specialization, in this already overcrowded labor market, they earned the lowest salaries, ended in low-level positions in federal and corporate bureaucracies, and
found fewer employment opportunities than in any other branch of engineering.
Although these bureaucratic jobs might have been demeaning for young men who
expected management positions, for women such positions offered relatively high
wages compared with other jobs available to them at the time. The drafting departments of State Highway Commissions gave temporary jobs to Knudsen in
Wisconsin and Elsie Eaves in Colorado during the 1920s, to Myra Cederquist in
Ohio in the 1930s, and to Emma Crabtree in Nevada in the 1940s.[31] The emerging
large corporations such as Westinghouse, General Electric, and Boeing also offered women an avenue to technical training through a kind of corporate apprenticeship. At Westinghouse, Bertha Lamme found ample opportunity to use her
superior mathematical knowledge and her engineering skills to design motors and
generators for over ten years, until she had to relinquish her job in 1905 when she
married a co-worker. Finding the door to engineering slightly ajar during the war
in 1917 as a young civil engineering graduate from the University of Michigan, Hazel Irene Quick established a long career, lasting until 1950, as a fundamental plan
engineer; she was in fact the only woman employed by the Michigan State Telephone Company.[32]

Even if the world wars offered opportunities to women, employers also responded to the modest increase in the number of women by setting up clear gender boundaries and by creating separate social and spatial arrangements. To deal
with the small increase of women, employers instituted sex-segregated offices and
drafting departments where some academically trained women could move into
supervisory but temporary positions. After graduating from Iowa State University
in civil engineering in 1894 and doing some graduate work at MIT, Alda Wilson
(b. 1873) worked in architectural firms in Chicago and New York for over ten
years, before she found a managerial job as superintendent of the women's drafting department at the Iowa Highway Commission in 1919. Unable to find an engineering position after the first World War, the overqualified and brilliant Edith
Clarke spent several years training and supervising women in the calculation of
mechanical stresses in turbines in a separate women's department within the Tur-
bine Engineering Department at General Electric in Schenectady, N.Y. These separate female spaces might have offered women a temporary niche but rarely a solid
stepping stone for full-fledged careers as designers, executives, or managers.[33]

Before the era of affirmative action, therefore, neither the federal government nor the corporations offered true alternatives to the kind of patriarchal patronage found by daughters of small firms. At the end of her career in 1947 when the government campaigned for women's return to their homes, the experienced Dennis wrote in the true spirit of belief in meritocracy: "we certainly do not want to discourage the ambitious young woman with the right qualifications for an engineering career," but she warned, "anyone pioneering in this field must be made to see that, outside of the lowest levels of clerical and manual work, there are almost no standard [management] jobs for women."[34] Dennis knew what she was talking about. Qualified women engineers and scientists of her generation had weathered the storm during the Depression in order to continue their careers. As Nora Blatch correctly observed when she worked as an engineering inspector for the Public Works Administrations in Connecticut and Rhode Island, federal sponsorships of women were limited during the New Deal. The investments of Roosevelt's public-work administration in major building programs provided engineering work for men only. The reforestation, highway, building, and reclamation projects were all closed to qualified women while they mobilized men at lower wages far away from their homes, leaving their wives to head their households. Moreover, the National Recovery Board still specified lower minimum wages for women than men.[35]

These highly educated women waited for better times and looked for jobs in teaching, drafting, or editing and secretarial positions with engineering firms and professional organizations. Strategically located as manager of the Business News Department at the journal *Engineering News-Record*, the Colorado graduate ('20) and socially well-versed Elsie Eaves provided mentoring and career guidance to many young women engineers during the 1930s and 1940s. She counselled them on how to get through the Depression, and encouraged them to acquire stenographic and secretarial skills in the hope of "a position with a fine engineer," but she warned, "I never encourage a girl to study engineering on the theory that if she wants it badly enough she will do it in spite of all discouragement."[36] During the second World War, when younger men went to the front and others moved up, women stood ready to take on the new jobs that were opening up, but never materialized. Instead of recruiting among the experienced women already available, the federal government chose to train young and inexperienced women. The state thus helped to institutionalize and intensify old patterns, as can best be illustrated by the politics surrounding the federal job of engineering aide.[37]

During the second World War, the federal job title "engineering aide" carried with it the heavy baggage of gender politics and amplified old habits. It defined women as non-engineers. As part of the war effort, American women like those elsewhere in Europe were encouraged for the first time to seek training in technical work. Under the auspices of the federal government and in cooperation with uni-

Figure 33. Three women draftsmen trained as engineering aides posing in classic Rosie-the-Riveter Government propaganda style in 1943. Photo campaign promoted by the Department of Labor's Woman's Bureau to advertise their importance to the war industry. Permission of Schlesinger Library, Radcliffe College, Cambridge, MA.

versities, large corporations urged young and bright women to apply for engineering jobs.[Figure 33] Federal agencies, large aircraft companies, and engineering schools pushed over 300,000 women through various kind of engineering programs ranging from three-month crash courses to college engineering curricula condensed into two years. Georgia Tech University in the South, like many other well-established schools hostile to any hint at coeducation before the war, opened its doors to women for a special training program sponsored by the U.S. Chemical Services when shortages of technically trained personnel threatened the war industries. The aircraft corporation Curtiss-Wright sponsored a course for women engineering students at several American universities including Iowa State, University of Minnesota, and Rensselaer Polytechnic.[Figure 34] The women received an engineering certificate after completion of the course which included work in engineering methods, mechanics, drafting, and processing.[Figure 35] Many of these specially trained women, who had been the best and the brightest in their high school and college, ended up in drafting, testing, and routine lab work, however.

Figure 34.   Women trainees posing for propaganda photograph for the Curtiss-Wright corporation's Cadette Program, an engineering crash course for women during the war in 1943. Courtesy of Archives of Women in Science and Engineering, Iowa State University.

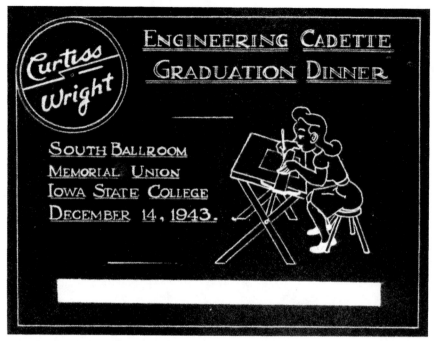

Figure 35. Drawing of a poster announcing dinner celebrating the graduates of the Curtiss-Wright Program. Courtesy of Archives of Women in Science and Engineering, Iowa State University.

Juliet K. Coyle trained in medical technology and biology as a young college student in 1943, when she was recruited by an aircraft company for a short course in engineering where she learned to read blueprints, drafting, statistics, and mechanical practice. When the war ended and she finished her studies, however, the company had no idea what to do with her and her equally well-trained women colleagues. The firm not only moved them around through different departments and paid them less than their male counterparts, but gave them explicit instructions to avoid giving orders to workers on the production floor – the male avenue to further promotion on the managerial ladder.[18]

Coyle's experience illustrates that none of these educational efforts, either during or after the war-time labor shortages, were meant to turn women into full-fledged engineers. [Figure 36] All programs were clearly intended for women's temporary employment as technical assistants to the various engineering departments, despite propaganda agencies' claims to the contrary.[19] During the second World War, vocational literature attempted to assure female recruits that such engineering work would lead to full careers, but Margaret Barnard Pickel, an adviser to graduate students at Columbia, questioned that promise and advised women to prepare for a backlash in peacetime. "Are the educators of women justified in en-

Figure 36.   Propaganda cartoons "Girls, Girls, Girls," as published by General Electric's Campus News, 1942-43.

couraging their students to start on the long, arduous and expensive training for an engineering degree with the expectation of a career at the end of it?" she asked. After taking an inventory of the barriers women would face, she concluded, "it seems hardly honest to hold out such a prospect as a professional possibility for women."[40] Dennis, often quoted by the Baltimore and Ohio Railroad for their public-relations literature during the war, also pointed out that "Women engineers have been ignored or else glamorized with newspaper publicity that is harmful to serious advancement in their work."[41] The Women's Bureau's investigation into the employment opportunities in peacetime was equally realistic and cautious, based on their experiences during the second World War. In explaining their research project, the officials at the Women's Bureau stated in their correspondence that "it is our opinion that the increase of women [in technical and scientific work] during the war has been greatly exaggerated because of the publicity presented to attract them. However, we want to find the facts through first-hand contact with professional organizations... those who employ women in technical and scientific jobs, and with training centers."[42]

Indeed, Margaret Pickel had been justified in sounding the alarm, as the Women's Bureau found. Tactics to constrain women – most of whom were college educated – took on various forms. Not only was a woman's job given the title of "engineering aide," but the implicit gender-based division of labor relegated women to drafting departments and laboratories, while men were assigned jobs on the production floor that enabled them to advance to managerial positions. In short, the federal policy created the term "engineering aide" to refer to women engineers, while it continued to use the title of "engineer" to denote men. The job title of "engineering aide" thus forcefully drew a line between technical expertise and management, and both reproduced and created a standard practice.[43]

In the U.S., as elsewhere, the relationship between management and the technical content of engineering as an occupation is critical to understanding the formulation of a male professional identity. Overqualified women found that the initial advantage of education would turn into a liability once they entered the workplace where male codes based on managerial qualities and hands-on experience determined one's professional standing. The gender division of engineering labor – between production floors and drafting offices – hinged on the very same (class) distinctions by which male engineers sought to distinguish themselves from self-trained foremen who had risen through the ranks. The division of labor between the laboratory and the drafting department on the one hand, and building sites and production floors on the other, became the single most important delineator between men and women in engineering. Wherever women engineers did succeed in gaining employment, they were most likely to be hired in drafting, calculating, or design departments, or laboratories and classrooms. In other words, women engineers joined the rank-and-file of the profession.

No woman without a family connection ever moved into supervisory positions in family firms. Women like Roebling, Gleason, and Gilbreth, who were steeped in the patriarchal culture of the family business advanced in engineering through a combination of excellence, perseverance, family connections, and the pooling of resources. But assessing the chances of women's employment in 1940, Olive Dennis (1880-1957), who like Blatch had received her education from Cornell University in civil engineering ('20) in addition to degrees in mathematics, warned that "unless a woman has a family connection in an engineering firm, or enough capital to go into business herself, her chances of rising to an executive position in structural engineering seem negative."[44] Employed first as a draftsman in the bridge division of the Baltimore & Ohio Railroad and later transferred to the company's service department for interior coach design, Dennis's response during the 1940s is the more revealing: She was always touted as a woman's success story, both by women engineering advocates and by her employer the Baltimore & Ohio Railroad, especially during the second World War, when the War Manpower Commission and the Office of War Information launched an intense propaganda campaign to lure women into the technical fields. Dennis's warning points to the split between government rhetoric and women's experience of it, the difference between women with family resources and those trying to make it on their own in the emerging corporate and federal bureaucracies, the gap between women's technical expertise and their ability to move into managerial positions, and the contrast between nineteenth-century ideals and twentieth-century practices.[15]

More explicitly than any woman might have said it, the introduction of the term "engineering aide" encapsulated the story of women's marginalization as a labor reserve force without the possibility for career advancement. With a single linguistic stroke, the term placed women with technical ability and training outside the emerging domain of technology.

FACING MALE PROFESSIONALISM

National professional organizations became the most visible if not the only institution of the engineering fraternity. Few scholars still regard the nineteenth-century movement of professionalization as a trend towards expertise, knowledge, rational behavior, peer review, and values void of any ideological concern. Most consider it a form of occupational control and autonomy with exclusive jurisdiction and privileges which conceal the advocacy behind the cloak of political disinterestedness and objectivity. The engineering societies were by no means exceptional; they explored some of the classic tactics pioneered by other professions, looking for new means to enhance their status and cultural authority. The classic model of professionalism defined by medical practice emphasized the work ethic, trust, professional associations, licensing, collegial control, and strong client and practitioner relationships was problematic for American engineers, as many historians of engineering professionalism have argued. In engineering the classic model also competed with business and rank-and-file models of professionalism, as Peter Meiksins has correctly argued. The American Institute of Mining and Metallurgical Engineers (AIME) advocated business values, closely associated with the culture of family firms, and adhered less strictly to the newly emerging ideology of professionalism. Its business-oriented policy had an immediate impact on the number of women admitted: while only 25 women had majored in mining by 1952 and many states had laws prohibiting women working underground, the mining engineers admitted more women to their ranks than any of the other major organizations. In 1943, the AIME membership included such daughters of family firms like Jean Horning Marburg, then member of the National Resources Planning Board, Helen A. Antonova, an assayer at the R&F Refining Co., Edith P. Meyer, a development engineer at Brush Beryllium Co., and another 19 women in addition to a large number of female students.[46] For daughters of family firms like Jean Marburg business professionalism opened some doors that would have remained closed otherwise. Thus, business models of professionalism were more open to women if they were connected to the patriarchal culture of family firms like Gleason, Gilbreth, and Marburg. They were closed to women without the proper family connections.

Even if engineers did not succeed to maintain strict professional boundaries compared to other professions when successful the classic model of professionalism turned out to be a thoroughly male and middle-class endeavor. The more an organization strove for professionalization modelled after the medical occupation, the more it was inclined to bar women.[47] It was not the number of women engineers in either absolute or relative terms which determined the percentage of female membership, but rather the level of professional claims to which the leadership of the respective national organizations aspired. Significantly more women engineering students opted for civil and electrical than other engineering fields, but this population was not reflected in the membership of the American Society of Civil Engineers and the American Institute of Electrical Engineers, whose requirements were more strict than those of their brothers in mining.

The American Society of Civil Engineers (ASCE) and other major societies guarded their boundaries against any female incursions. Other organizations also granted secondary membership without voting rights to unimportant rank-and-file engineers and women who managed to infiltrate them.[48] The American Institute of Electrical Engineers (AIEE), emerging in a field with high aspirations towards a medical model of professionalization, refused to admit Susan B. Leiter, a lab assistant at the Testing Bureau in New York, to membership in 1904, when the organization was looking for ways to upgrade the profession.[49] When Elmina Wilson and Nora Blatch applied for membership to the American Society of Civil Engineers, they found the doors closed. With a *cum laude* in civil engineering from Cornell University, Nora Blatch could claim superior mathematical ability and theoretical engineering knowledge, the kind of credentials advocates of engineering schools thought crucial for any engineer to succeed. But Blatch had more to offer: she also possessed the necessary hands-on experience that advocates of shop floor and field training saw as hallmarks of the true engineer. These was all to no avail. In 1916, when she turned 32 and the society dropped her from its membership, Nora Stanton Blatch filed a lawsuit against the ASCE. As an experienced engineer, she had accumulated over ten years of experience to meet the Society's requirements. In addition to her four-year education in civil engineering at Cornell, she had taken courses in electricity and mathematics with Michael Pupin at Columbia University. She had practiced as a draftsman for the American Bridge Company and the New York City Board of Water for about two years and as an assistant engineer and chief draftsman at the Radley Steel Construction Company for another three years and finally as an assistant engineer at the New York Public Service Commission. Most importantly for the requirements for full membership, she had supervised over thirty draftsmen when working at Radley Steel. Blatch, a feminist, divorcee, and single mother whose income depended on her engineering work at the time, challenged the ASCE when more women and sons of lower-class men were trying to enter the field through the new institutions of higher education and when engineering advocates were busy defining the occupation as a profession by excluding more and more groups of practitioners such as draftsmen and surveyors. Her suit marks one of many contests in which the emerging professions staked out their professional claims by means of border disputes with other competing fields.[50]

In addition to outright exclusion, the professional organization dealt with what they saw as female incursions through the tactic of granting women secondary membership without voting rights. The controversy over Ethel Ricker's Tau Beta Pi membership marked yet another drawn-out contest over gender boundaries in the fields of civil engineering and architecture where women were numerous. In 1903, the local chapter of Tau Beta Pi elected Ricker, an architecture student at the University of Illinois, to the engineering honor society, but the national executive board and the society's convention not only overturned the decision to elect her, but went so far as to amend its constitution to specify that henceforth only men would be eligible for membership. During the 1930s depression, when many civil engineers faced unemployment, the Tau Beta Pi honor society introduced a Women's Badge in an attempt to deal with the (small) number of qualified women who had made their presence felt. To make it clear that their acceptance should be considered an act of benevolence, Women's Badge wearers were neither members nor allowed to pay initiation fees. Disapproving of such "separate but unequal" recognition, some women refused them. Nonetheless, it would take three-quarters of a century before women would be accepted as equal partners in the organization: in 1969 the honor society changed its constitution to admit women, in 1973 it sanitized the constitution and bylaws of sexist language, and in 1976 it elected a woman as a national officer for the first time. By changing its constitution and by designing a "woman's badge," the fraternity of young aspiring engineers set up explicit gender barriers around engineering when job markets were particularly tight. The resistance of Tau Beta Pi resulted from a mutual shaping process between the newly established specialization's need for social and professional status and the increasing numbers of women architects and civil engineers demanding their rightful place behind drawing boards and on building sites.[51]

Like the various scientific fields, engineering specializations were sanitized from the perceived threat of feminization either by excluding women from full membership in professional organizations or by relegating them to a secondary status without voting rights. The development of subordinate, segregated female professional cultures in areas such as chemistry, however, proved to be the most crucial tactic to the very definition of a profession. In the words of Margaret Rossiter, "the very word *professional* was in some contexts a synonym for an all-masculine and so high-status organization."[52] Women faced outright exclusion, were relegated to secondary membership, and banished to separate organizations.

Divide and Conquer

The definition of separate labor markets for male and female engineering work presented another tactic to deal with women who started to seek engineering education and employment. The best example comes from the chemical industry where many women found employ as chemists. To deal with the female incursions the American Institute of Chemical Engineering did everything in its power to define the occupation in such a way as to effectively bar women from the field and relegate them to chemistry.

The ability to manage other men became the key to the chemical engineers' definition of their profession in an effort to distinguish themselves from chemical analysts – a large proportion of whom were women – whose status and pay diminished dramatically around 1900. In response, production chemists sought to align their occupation with the male world of mechanical engineering rather than with the world of science. Chemical engineers saw themselves as running plants, as opposed to labs. As one of the most important founders of The American Institute of Chemical Engineers (AIChE), Arthur D. Little (1863-1935), spokesman of engineering professionalism and the nestor of commercial chemical research, introduced a key concept for the development of a distinct chemical engineering identity in 1915: the notion of "unit operations." Little argued that unit operations involved neither pure chemical science nor mechanical engineering, but distinctly physical, man-made objects in the plant operation rather than chemical reactions in the lab. In the same year that Little refined his notion of chemical engineering in a report for the AIChE, his opinion was solicited by the Bureau of Vocational Information for a report on employment opportunities for women in chemistry and chemical engineering in the tight engineering labor market after the first World War. Recommending chemical analysis as "one of the most promising fields of work for women," Little reserved chemical engineering as an exclusive specialization for men, arguing that "it is probably the most difficult branch of the profession." In addition to the long hours, the extensive travel, and the physical endurance needed, he considered the "rough and tumble of contests with contractors and labor unions," involved in the new construction and design of a plant, to be prohibitive factors.[53]

Little's rhetorical position was broadly shared by male chemists working in the field. While stressing women's strength in all other lines of work connected with the chemistry lab, the chief chemist of the Calco Chemical Company voiced the general sentiment in 1919: "It is impossible to use women chemists on development work which has to be translated into plant practice by actual operation in the plant. This is the only limitation." Other potential employers of women chemists elaborated on that particular theme by explaining that "research men must go into the Plant and manipulate all sorts of plant apparatus, direction [sic] foreign labor of every sort. You can readily see that a woman would be at a great disadvantage in this work," or that, "it often involves night work and almost always involves dealing with plant foremen and operators not easy to deal with." The work of chemical engineering involved "large rough mechanical apparatus... which work is usually carried on by unintelligent labor, in a good many cases the roughest kind of material," the representative of the Grasselli Chemical Company's research department wrote in 1917.[54] By establishing such boundaries, Little and other chemists succeeded in safely associating their work with both the male codes of the machine shop or the plant operation and managerial control since the power struggle in the workplace where matters *of* class were contested was a matter *between* men.

Despite these ideological constructs of women's and men's engineering work, it is questionable whether women did not, in fact, do such work. Women chemists were sometimes responsible for tasks bordering on chemical engineering, but they were rarely upgraded to that level. Often discrepancies existed between job title and job content. Take the case of Glenola Behling Rose, whose job title was chemist but who described her duties in 1920 as follows: "I left the chemical dept. to go into the Dyestuffs Sales Dept. I have but one man over me and as his assistant, I am the *Executive Office Supervisor* of the Dyestuffs Technical Laboratory and have charge of all dealings with the chemical dept. such as deciding what dyes they shall go ahead to investigate & in what quantities, and keep track of their work in order to see whether they produce the dyes economically enough for us to market them. In a way I am the link between the research, the manufacturing and the selling of dyestuffs... As you will see a good deal of my work is supervisory." With Bachelors degrees in geology and chemistry, and a Masters in chemistry, the highly qualified

Glenola Rose felt she was technically well-prepared for such a job. In response to the question of what training she thought would be most beneficial to women entering her field, she replied that there was a "need for a thorough foundation and a training with men," by which she meant the task of managing men. And Florence Renick wrote that, in fact, "I have had to deal considerably with labor of all kinds, mostly ignorant and many foriegners [sic] among them, and none of them but consider me 'boss' so far as the laboratory is concerned." On this particular point Jessie Elizabeth Minor, chief chemist at the Hamersley Manufacturing Company, articulated women's ambivalence: "There is still much masculine prejudice to combat. Many laboratories are not attractive looking. We come in contact with working men (which may be construed as an asset or liability)."[55] Thus in these contexts, white women chemists actually did supervisory work and would have qualified as engineers according to the terms Little and other chemical engineering advocates had established for their engineering specialization. In all these instances, technical qualification or experience was less decisive in considerations for job assignments and promotions than the issue of supervision.

Women chemists were thus kept under job titles they had actually outgrown. Significantly, traffic between chemistry and chemical engineering also went in the other direction. Women who trained as chemical engineers ended up in lower-paid positions as chemists. Dorothy Hall (1894-1989) might have been the success story of a woman advancing on the corporate ladder at GE as a research and later chief chemist, but with a Ph.D. in chemical engineering (University of Michigan '20) she was overqualified for her job. When asked, most employers said they thought women competent and excellent for research and analysis; few raised objections of a technical nature. But all drew the line at work related to the plant operation: they stressed, as had Arthur Little, labor-related issues. Indeed, in 1948 the Women's Bureau reported that most women who trained in chemical engineering were employed as chemists.[56]

These examples are telling indeed. Large numbers of women interested in the sciences in the U.S. and elsewhere flocked to the field of chemistry. The same held true for chemical engineering: more women graduated in chemical engineering than in any of the other engineering specializations. In fact, a higher proportion of female engineering students than of male engineering students majored in chemical engineering. In a limited way, chemical engineering offered a niche to women students interested in engineering. Prominent chemical engineering advocates pushed for an explicitly male professional ethic by defining their discipline as an exclusively male domain which required supervisory skills. Tens of thousands of women chemists and chemical engineers were banished to chemical labs, where working conditions were dire and the pay scales low.[57] In these contexts, a chemist meant being an ill-paid woman while a chemical engineer often denoted a man in command of higher wages and managerial authority.

The male establishment engineers' push for professionalization was in part a response to the enormous expansion of engineering work, which provided new opportunities to lower-class youths and sons of recent immigrants. The call for clear boundaries of class, however, resulted in the kind of male middle-class reinscription. The census data further symbolized the process of exclusion and reinscription articulated in these linguistic boundary wars. In their unceasing efforts to find new categories for reliable enumeration, census takers continued to look for a standard of consistent classification by excluding more and more groups of skilled workers from the category of engineers, including boat and steam shovel engineers, foremen of radio stations, engineers under 35 without a college education, and chemists. According to economists who have worked with the data, however, the statisticians made these adjustments without much success in terms of uniformity. These statistical and linguistic interventions did little to generate a satisfying set of data.[58] More broadly speaking, these interventions were part of the process of professionalization in which men like Arthur Little engaged. Nevertheless, historians have reproduced many of these figures, including the definitional exclusions they represent. The definition of who would count as a true engineer and the production of statistics to justify this illusion mutually shaped each other. More generally, the example of the chemical industry shows the kind of linguistic constructs and social practices involved in making women invisible as engineers.

ORGANIZING AT LAST

Women engineers responded to such tactics with stoicism but also with collective action. Before the first World War, the early generation of women engineers like Richter, Wilson, Blatch, and Leiter had tried to gain access to the existing male organizations as individuals. They were rebuffed outright, granted secondary status, relegated to separate-but-unequal organizations, or segregated into different labor markets. A second generation of young women students and recent graduates including Lou Alta Melton, Hazel Quick, Elsie Eaves, Hilda Counts Edgecomb, and Alice Goff, who had found entry-level employment opportunities during the first World War, tried in 1919 to organize collectively into a separate women's organization, but failed. The post-suffrage generation championed the cause of women engineers with great enthusiasm. Yet, at least publicly, none of the women's advocates rallied to the feminist cause, even if they grounded their promotion of women engineers as professionals precisely in one of the important principles of modern feminism: as individuals, women should be able to develop themselves to their fullest potential. All supported the notion that women had the freedom to choose whichever line of work suited their abilities, without the obligation to appeal to feminine propriety by arguing that such a choice was inspired by higher morals. All ardently believed the engineering profession's promise of upward mobility. Resisting any direct association with the women's movement, they claimed instead that they just happened to have a knack for engineering – a kind of discourse particularly dominant in 1943 when government propaganda sought arguments to mobilize women for the war industry.[59] The majority of women engineers then believed in, and had internalized, the values of corporate engineering, merit, and self-reliance.

The second generation saw their organizing efforts thwarted partly because, in search of much-needed recognition, they tried to keep up with emerging high professional standards by excluding from membership engineering students and working women engineers without formal education like non-collegiate draftsmen, chemists, and testing technicians.[60] They did so no doubt in an effort to defend against sexism and to garner greater prestige, but emulating high professional standards prevented them from gathering the critical mass necessary for such an organization.

In the same year that American women tried to organize, their British colleagues succeeded. British women established the Women's Engineering Society (WES), an inclusive organization that encompassed women engineers with or without formal collegiate education as well as machinists who were skilled or semiskilled workers. The British successfully, albeit briefly, united across classes in part because they did not adhere to the classic model of medical professionalism but to the tradition of high-class business professionalism and trade associations combining them with feminist ideals. In the end, the British women leaders too abandoned their policy of "gender solidarity for male privilege and class advantage" and narrowed "their focus to exclude the great mass of women who had entered the engineering trades during the first World War directly out of the working class," as its historian Pursell has argued.[61] Between the world wars, when job opportunities virtually disappeared, American women engineers sought temporary shelter with their British colleagues through membership in the WES and kept in touch through informal networks. Clinging to the medical model of professionalism, but failing to gather a critical mass, American women engineers of the interwar period turned instead to the tactic of trying to shape public opinion and writing biographical sketches of each other according to well-established formulae which stressed that with hard work and self-reliance women could indeed become engineers, to paraphrase Alice Goff's publication during these years. American women engineers borrowed from male models of merit, but neither questioned the structure of engineering that fundamentally hampered their chances nor campaigned for equal rights.

The final push towards organization in the U.S. did not come from the hundreds of thousands of women working in federal engineering jobs during the second World War, from the informal networks of academically overqualified women engineers who had learned to be stoic and self-effacing during hard times or from the women urban planners who had been nurtured by the women's reform movement. It came once again – as it had in 1919 – from young students and recent graduates eager to enter the job market and yearning for official recognition and respectability.

After the gap between rhetoric and reality had widened once again, American women engineers united at last in 1949. They did so long after other female professionals like lawyers and doctors had succeeded. An energetic and ambitious junior student leader on a scholarship, Phyllis Evans, won the support of Dean of Women Dorothy R. Young and university counsel A.W. Grosvernor to organize the first meetings of women engineers at Drexel University in 1949.[Figure 37] She and her colleagues organized over seventy young women engineering students from 19 colleges on the East coast to have their "voices... heard in the technological world," to address inequities in engineering work. In the greater New York area a group of women engineers who had been working in war-related industries – students from Cooper Union and City College of New York coupled with graduates working in the area – also struck up conversations about their plight with their sisters in the college libraries and Manhattan's coffee shops. Soon the long hidden tensions over leadership and the direction of women engineers' professionalism burst onto the scene.[62][Figure 38]

In the founding years, the student group at Drexel University in Philadelphia and a coalition of various groups in the greater New York area were in competition with each other for leadership. Its origins can be traced to a contest over the different professional strategies. Phyllis Evans – like so many other young women who had begun studying engineering during the war – was not yet married, about to graduate, and facing unemployment. Echoing the governmental war rhetoric, she cherished high expectations for her future. Explaining her choice for engineering, she told a journalist of her war experience as a cadet sergeant that had inspired her to go into engineering and her hopes for the future in military research: "I want to build rockets and I want to go to Mars," she said with youthful optimism. Establishment engineers like Lillian G. Murad and Lillian Gilbreth, propriety owners who were steeped in the ethics of the patriarchal culture of family firms, opted for a more conservative strategy involving supervision and high professional standards. Although Lillian Gilbreth had been supportive of women engineers, she did not favor a separate women's organization and was disinclined to head the SWE when it first looked for leadership. She was most concerned about the bold approach of a separate organization and its feminist implication and warned against

Figure 37.   Photograph of Phyllis Evans posing in overalls in Rosie-the-Riveter iconography illustrating a newspaper report on the first organizing efforts of women engineering graduates in *The Christian Science Monitor* (April 19, 1949). Courtesy of Royal Dutch Library, The Hague, The Netherlands.

blaming men for the difficulties women encountered in entering the field; instead she accused women of a lack of requirements. "The reason for women not being admitted into the National Engineering Society was not because they were women, but rather because they did not yet meet the qualification," she said. But the elder Gilbreth was somewhat at odds with Dorothy Young, Drexel's Dean of Women's students, who pushed for an activist strategy that confronted the inequity between men and women, while also striking a conciliatory note: women "need to realize that it is necessary to work cooperatively with men in larger field, planning together to abolish those inconsistencies that mar our democratic society".[63]

Figure 38.   Group portrait of women attending the founding meeting of the Society of Women
Engineers at Green Engineering Camp of Cooper Unions, New Jersey, May 27, 1950, as a response to the government's back-home campaigns just after the war. Courtesy of Archives of Labor and Urban Affairs, Wayne State University, Detroit, MI.

For the young organization, its strategy remained a balancing act between the
impatience of the younger women students of Miller's generation whose expectations had been raised by the government propaganda of Rosie-the-Riveter and the
cautious but conservative strategy of a previous generation of daughters and of
wives wedded to the patriarchal culture of family firms. The SWE tried to inspire
younger women to go into engineering as a promising career path by using individual women's careers as role models. Through the establishment of award med-
als and scholarship programs, biographical narratives, and pictures, the
organization stressed lone individual efforts rather than the new corporate male
ideal of team players propagated by corporations like Dupont and General Motors. It sponsored merit and self-reliance rather than a collective movement battling inequality for which Blatch campaigned.

The SWE never resolved these conflicting goals. It sought to attract more
young women into engineering schools yet necessarily had to deal with the
long-entrenched patriarchy of family firms, the tactics of the professional organi-

Figure 39.   Part of the campaign to recruit women for engineering positions for the U.S. military as the Cold War heats up in 1953. Gender relations are preserved rather than subverted by the photographer's frog-eyed frame stressing masculine features of the woman in charge and the 'subordinate' position of her reclining colleague. Permission of Schlesinger Library, Radcliffe College, Cambridge, MA.

zations, the discriminatory employment practices of corporations and the federal government. To the extent that it openly battled prejudice and sexism in engineering schools and practice, however, it risked frightening off prospective recruits.[64] If the SWE had been established in 1949 in the postwar period of "adjustment" when government and corporations devised policies to push women back into their homes, the organization would ride a new wave of ambiguous government encouragement during the Cold War.[Figure 39] Like the former Soviet Union, the former German Democratic Republic, and other Eastern European countries, the American military collaborated with corporations in actively recruiting women as technical personnel and initiated its "Woman Power" campaigns after the Sputnik panic in the West in 1957. Despite its evocative title *Woman Power*, a 1953 rapport

Figure 40. General Electric's advertising promoting the company through four of its women engineering employees against a male-coded apparatus. The image countered the working-class image of Rosie the Riveter of the government war propaganda campaigns to show women could be engineers without losing their femininity during the Cold War. Reproduced from Cincinatti News 1959.

issued by the National Manpower Council at Columbia University had little to do with feminist calls for equal rights: it shied away from controversial issues like equal pay for equal work and job discrimination, for which Nora Blatch had campaigned throughout her life. It neither upset gender hierarchies nor helped foster a separate women's culture. Instead, it did provide a certain legitimacy to a conservative part of the women's movement stressing that women, if they worked hard, could labor for the needs arising out of national shortages.[65] The new society successfully gathered the critical mass necessary for such a separate organization despite its high professional ideals because of the cold-war industry's demand for technical personnel that had been lacking three decades earlier in 1919. The truth was that women engineers' individual opportunities were part of America's military-industrial complex and highly depended on it. At times, its doors might be open to women, but it almost always reproduced old patriarchal patterns in a new corporate context.[Figure 40]

"Woman Power" and Daughters of Martha: Failed Allegories

The government's campaign of *Woman Power* was a borrowed identity. So was Lillian Gilbreth's appropriation of Rudyard Kipling's 1907 poem "The Sons of Martha."[66] At the 1961 opening of the SWE's headquarters in the United Engineering Building, Gilbreth tailored the poem to the needs of women engineers by entitling her speech "The Daughters of Martha." The modernist United Engineering Building towered high in New York and expressed the coming of age of the engineering professions, but the new headquarters Gilbreth was about to open were tiny and symbolic for women's place in the profession. Her reworking of Kipling's allegory showed the narrow place of women within the male world. Gilbreth sought to empower women in the technical professions by invoking the specter of service, sacrifice, and self-reliance. It was an unintentional but correct assessment of the double hardship of women who labored on the lower rungs of the profession as rank-and-file engineers and corporate workers. She thus validated a model that doubled the burden of women who aspired to be engineers: women engineers were expected to make sacrifices by virtue of both their gender and their profession.[67] By the 1960s, Gilbreth's professional strategy of "simple service simply given," as she cited Kipling, or as she had advised earlier, "helping others express themselves [as] the truest self expression," was out of date for women.[68] Her 1961 call for the inclusion of women in the profession was based on her own long career, on the trying experiences of many other women engineers, but principally on a conservative strategy of stoicism and overqualification for women professionals in general. Despite her attempt to redefine the place of women in engineering, the celebration of service, sacrifice, and self-reliance also reinforced some very traditional notions about women's proper place in engineering. Her employment of Kipling's poem threatened to become a failed allegory. When male engineers used the poem, it could be mobilized to appropriate working-class badges of manliness or to symbolize them as underdogs, but when women mobilized the poem for their cause the figure of Martha turned into an image of subordination, stoicism, and lack of advancement. Men perhaps could pass "down," but women could rarely pass "up" the cultural hierarchies as women found here and elsewhere.

Women engineers in the military-industrial complex or in the patriarchal culture of family firms had no appealing role models except the very ambiguous image that Lillian Gilbreth supplied. Outside the military-industrial complex and the patriarchal family firms, women were building their own structures: the women of the Progressive era who participated in the women's reform movement helped shape an alternative women's technical culture that was nurtured by women's traditions.[69] Throughout the country from Boston to San Francisco, women reformers helped build the public infrastructures in the civic improvements movement as private citizens rather than as corporate employees. Highly organized in private philantrophic organizations like the General Federation of Women's Clubs, these women reformers effectively became "municipal" housekeepers of the world. They conducted surveys, drew up plans for urban infrastructures, pushed for better housing, and helped finance public facilities from streetlights to sewer systems. They forged coalitions with local politicians, architects, civic leaders, and professional women like Ellen Swallow Richards, Alice Hamilton, and Ruth Carson in public health, science, and social research. As historian and suffragist Mary Beard demonstrated in 1915 in *Woman's Work in the Municipalities*, these women of the Progressive civic movement became in fact urban planners of the modern age.[70] This rich female heritage of building was not available to women engineers working in family firms or military and corporate industries. In fact, the women of these separate technical cultures never met or built bridges between them. More importantly, perhaps, the kind of distinct female building tradition Beard traced fell outside the modern but recent definitions of engineering and technology.

EPILOGUE

Gender, Technology, and Man the Maker

T he intellectual constructs and material practices of technology discussed in this book came to produce the world of industrial capitalism. Modern meanings of technology arose from the convergence of discourses around a number of nineteenth-century terms related to the rise of industrial capitalism, most importantly the rhetorical positions about *useful arts, inventiveness,* and the *machine.* Each of these terms was the focus of struggles in which middle-class professional men – among them engineers – staked their claims on key aspects of industrial capitalism, to the exclusion of women, African-Americans, and workers. In the U.S., these discourses merged in the 1930s, first through the work of Thorstein Veblen, who turned technology into an all-embracing concept linking it to engineers rather than workers, and through the debates over technocracy. By the late 1930s, technology had become firmly linked to the male, middle-class world of establishment engineering. Signalling the success of that century-long ascendancy, in 1978 *Encyclopaedia Britannica* designated engineers as the true bearers of that knowledge in its first entry on technology. Dictionaries and encyclopedias have always been more than descriptive texts prescribing and shaping new categories of thought and recapturing old ones. Operating in a larger political universe, they canonize the dominant discourse through their choices of what to include or exclude from their pages. The 1978 edition of the *Encyclopaedia Britannica* included an entry on "Technology" for the first time, stating that "[b]y virtue of his nature as a toolmaker, man is therefore a technologist from the beginning, and the history of technology encompasses the whole evolution of man."[1] The *Britannica* canonized the term as late as 1978 because it is only in the twentieth century that we encounter the idea of technology in its present popular form: as an autonomous factor, an artifact, a fact, but also as an essentially pristine and neutral form of reality that requires little explanation. Engineers were designated as the sole bearers of this form of knowledge. *Technology* as charted by the Encyclopedia functioned as a kind of narrative strategy that includes both intellectual constructs and material practices.

Historian, author, and suffragist Mary R. Beard (1876-1958) confronted Britannica's assumptions and challenged its male genealogy. The *Encyclopaedia Britan-*

nica, which despite its name had come under American intellectual and financial control in the early twentieth century, claimed inclusiveness and objectivity. The mantle of cultural authority assumed by the encyclopedia did not go unchallenged. From the very start, *Britannica*'s slogan, "All life in one set of books – the richness of the human mind, the achievements of all civilizations, the problems and hopes of the future", made it vulnerable to accusations of exclusion and distortion.' Understanding the importance of the *Britannica*'s growing cultural authority in the first part of the twentieth century, Beard tried to (re)insert women and their contributions into the canon. After her activism in unions and the suffrage movement, she turned to a life of writing and lecturing. From her early book *Woman's Work in Municipalities* (1915) to her most famous book *Woman as a Force in History* (1946), Beard's main goal in life was to set the historical record straight by arguing that women were a powerful force in the creation of civilization and culture and that a history without them would be incomplete. Among her many activities as a feminist historian, Beard first campaigned for a World Center for Women's Archives. When the initiative failed, she and Marjorie White opted for an encyclopedia of women in the 1940s, and began to contest past encyclopedias' claims of all-inclusiveness. In response to her attack, the *Britannica*'s editor Walter A. Yust invited her to make suggestions. Beard and several women researchers responded to his overture by offering some corrections and by attempting to rectify the Britannica's overall male taxonomy and bias. Beard pointed not only to the many women who should have been included, but described a pattern of sexual differentiation over time and suggested how old materials should be reinterpreted anew. She also showed the mechanisms that reduced and excluded women's roles in modern civilizations.

The work of Gage, Morgan and Mason influenced Mary Beard when she argued that the entry on *Labor, Primitive* was among the weakest of the articles. It ignored the "story of woman's original creativeness as inventor of the industries, arts, and as the first farmer by reason of her concern with soil – not her slavery." Along the same lines, she criticized the entry on *Weapons, Primitive*, suggesting that it ought to be called *Social Implements* instead. "They [the weapons] are reviewed as features of the male's fighting function or inclination. But primitive weapons for social construction are not even hinted at. Surely they were even more important." Echoing Gage and Mason, Beard suggested "introducing an article on constructive implements (not calling them weapons), beginning with a section on *Primitives* and frankly acknowledging the leadership [of women] in this connection as inventors of cooking (with implements), spinning, weaving, etc."' Despite the invitation of the *Britannica*, Beard and her co-workers' two years of hard work came to nothing. Instead of revising the very building blocks of the *Britannica*, its editor chose to include some biographical sketches of notable women. It did little to challenge the encyclopedia's parameters and only reinforced women's margin-ality.[4] As *Britannica*'s most recent entry on technology illustrates, Beard's critique remains as valid today as it was fifty years ago.

The story of the entry on technology in the fifteenth edition of the *Encyclopaedia Britannica* in 1978 epitomizes the paradigmatic shift in the understanding of technology over the last two hundred years. It is emblematic for the formation of the new terminology that we have come to accept it as a natural male domain defy-ing any explanation. The *Britannica* contained no entry for technology before the supplemental 14th edition in 1968, yet by the completely revised 15th edition in 1978, technology had been elevated to a central position. The *Britannica* was not alone in the late admission of technology to its pages: other encyclopedias likewise neglected inclusion of the category well into the 1960s. Alongside such other cate-gories including Matter and Energy, the Earth, Life on Earth, Human Life, Hu-man Society, Art, Religion, the History of Mankind, and the Branches of Knowledge, *technology* now assumed center stage as one of the ten organizing prin-ciples of knowledge.[5] Significantly, *Britannica*'s new entry replaced the one on technical schools, whose engineering educators had first introduced the word to the English language to stake out a new area of expertise; it was absorbed within the narrative of technology.

By 1970, the term had migrated out from the walls of engineering schools to en-ter the world at large to appropriate a kind of objectivity. But its claim went fur-ther. If its new lemma *Technology* encapsulated the old one on *Engineering Schools*, the claim was even much more dramatic, for it also bid for the entire expanse of the human universe. As *Technology* was promoted from sub-entry to main entry, so its authors, Eugene Ferguson (1917) and Melvin Kranzberg (1916-1996), were pro-moted to the higher echelons of the editorial board. Ferguson had trained as an en-gineer and had worked for the chemical corporation Dupont in Wilmington, Delaware, from 1938 until 1942, before teaching the history of technology; for Kranzberg, his contacts with the military during the second World War and the Stevens Institute of Technology proved to be seminal in shaping his understand-ing of technology. After receiving a PhD in French history from Harvard in 1942, Kranzberg enlisted in the military, where he trained in electrical engineering at Catholic University, Johns Hopkins, and Philco Radio Laboratories. While teach-ing at the Stevens Institute of Technology in 1946 and 1947, he sought to bring his-tory closer to the experiences of his engineering students. Both Ferguson and Kranzberg were movers and shapers of America's first professional organization, The Society of the History of Technology, and its home journal *Technology and Culture*, in the late Fifties.[6]

While Kranzberg and Ferguson were still debating the concept, approach, and material for the entry with each other, a chief editor of the *Encyclopaedia Britan-*

nica informed the writers, "[w]e have finally come to believe that we cannot, after all, push technology under the big 'arts' tent, no matter how broadly 'arts' is defined. You have, therefore, been elevated from *Division* to *Part* status."[7] With these words, the editor congratulated Ferguson and Kranzberg on their new job description. He promoted them from the subordinate position of a sub-entry, or *Division* of the Encyclopaedia, to one of the ten slices, or *Parts*, of what the chief editors called "The Circle of Learning." In proposing the image of the circle, the editors sought to avoid any implications about a hierarchical order or unilinear and evolutionary development of knowledge.[8] That, at least, signalled the intention, but as classifiers and encyclopedias are wont to do, the editors created hierarchies of knowledge, producing a narrative of their own. The authors charted a new course for technology: the entry first described prehistoric times according to archaeological classifications, which measured periods in terms of the materials and the tools found in excavations including stone, fire, bow and arrow, potter's wheel, and building structures. Throughout the entry, the writers emphasized the beneficial relationship between science and technology, which they believed to be part of the development of technology.[9] Despite the considerable attention devoted to prehistoric times, the Middle Ages, and to a period called "The Emergence of Western Technology (1500-1750)," the Industrial Revolution assumed center stage. In their discussion of technological developments in agriculture, the authors emphasized the British enclosure movement as a prelude to the Industrial Revolution and neglected the groups dislocated by these movements. The types of tools and devices discussed also reflected this bias. About a quarter to a third of the entry dealt with tools that skilled and unskilled workers designed, produced, or handled. By contrast, the majority of the machines and devices listed belonged to engineering.[10] In general, the entry omitted all groups opposed to and displaced by such developments as well as the roads-not-taken in the development of technologies. Few if any of the technologies discussed had any association with women's work. Moreover, the writers often removed social actors from the story, describing instead the manufacturing processes involved. In a short concluding section, they sought to correct and justify some of their choices, for example, by devoting special attention to the science and technology relationship, the intellectual protests against technology, its ecological costs and destructive side effects, and its potential for "underdeveloped regions." The writers defended the inherent inevitability of technology, for example, by asserting that Gandhi might have hoped for a return to the pre-industrial economy symbolized by the spinning wheel, but that this idea was hardly shared by his countrymen who embraced modern technology.

Kranzberg and Ferguson also wrestled with the new importance accorded to machines that had been so popular with visual artists, public intellectuals, and engineering and science spokesmen like Duchamp, Veblen, and Mumford. They recognized that the traditional artifactual approach to what they referred to the "taxonomy of machines" hampered rather than helped their project. As Ferguson explained, "[b]ecause I have been familiar with earlier attempts to classify and anatomize machines, I have been quite skeptical of finding any particular sensible approach since 'machines' can only be anatomized…if they are restricted to an assemblage of linkages." He especially questioned the methodology of the German academic engineer and writer Franz Reuleaux (1829-1905), whose hierarchical classification system attempted to group machines by their internal structure without looking at the environment in which they had been produced and used." Ferguson objected in particular to the emphasis on formal and academic knowledge of American mechanical engineering education over that of tacit knowledge, but at this time he did not follow up on this insight in the entry." More importantly, the authors sought to distance themselves from the idea that technology could be carried on autonomously without human agency. In the final analysis, the writers asserted that man remained the master and the center.

The authors did not question the new centrality of engineers, stressing that "better educated scientists, engineers are needed both to operate and to criticize the increasingly complex technological apparatus." As a matter of fact, they did much to single out and bring engineers, scientists, and inventors to the fore, not shying away from somewhat novel and even anachronistic classifications to attain that goal. They charted a new historiography in their choice of privileging mechanical and civil engineering and departing from earlier genealogies. In an effort to place the history of technology on an engineering base, they claimed that the cotton and textile industry's "importance in the history of technology should not be exaggerated," despite their acknowledgment that it "probably more than any other, gave its character to the British Industrial Revolution." Whether intentionally or not, the shift in historiographic focus from the textile and cotton industries to engineering meant that both women and workers lost visibility. These choices rendered the technological domain even more exclusively male-coded than before.

The 1978 *Britannica*'s entry for technology is emblematic in many ways. Its late appearance indicates that as an intellectual construct and material practice the concept belonged to the post-World War cultural practice. Despite its late entry within the encyclopedic production, when it finally appeared it was as swift as it was profound. The editorial board's decision to elevate technology from its status as a sub-entry to one of the ten wedges in the circle of human knowledge reveals in a single incident the enormous importance that the term and concept acquired within just a few decades. While the idea took two centuries to ripen, the prominence of technology as a keyword was sudden and far-reaching. By 1978, technology had become the most important diagnostic tool to measure human worth. Most remarkable to historians of technology, the *Britannica*'s wheel of human knowledge excluded science. Scientific knowledge and research still found its way into many entries, but science was neither a constituent element nor an organizing principle of the *Britannica*. Technology now performed this role. This was perhaps not such a radical departure as it first seemed. As spokesmen for technology, engineering advocates from Thurston to Karl Compton, Robert Millikan, and Arthur Little had extricated the profession from its labor associations and shop-floor knowledge by appropriating the idiom and authority of science. This new discourse was subsumed into the discourse of technology. By extension, the *Britannica* entry may be seen as a triumphant manifesto of technologists' independence from the attempts of pure science advocates to relegate the useful arts to an inferior place. If Science – in the sense of a capitalized, corporate, commodified, theatrical, and exhibitionist kind of science as Americans experienced it at World's Fairs – developed a language through which technology came to be understood in the nineteenth century, the reverse held true for the twentieth century.[1] Science and technology had long played musical chairs in their quest for cultural authority, but when technology suddenly rushed on the American scene as a keyword during the 1940s, it was far reaching and enduring.

The *Britannica*'s entry on technology represents at once a description of the technological development, a recapitulation of the scholarship, and an elaboration of an emerging popular understanding of the concept. Kranzberg and Ferguson reformulated and formalized a discourse classified as technology that by 1978 embraced the entire span of human evolution. While Lewis Morgan had argued that inventions pushed forth human evolution in the nineteenth century, technology now represented the very substance of human evolution. In addition, the entry had a programmatic intent, because the writers wanted to engineer a new program for an emerging discipline, the history of technology: they placed engineering at its center.[14] This was no coincidence. Ferguson, like many other pioneering historians of technology, was a lapsed engineer in search of status for his profession by providing it with firm antecedents in history while Kranzberg was one of those historians who worked at engineering programs and looked for novel approaches to teaching the past in ways that would interest undergraduates.[15]

Ferguson and Kranzberg's *Britannica* project illustrates a laudable attempt to avoid the pitfalls of a machine-bound understanding of technology that had captured the literary, visual, and public imagination before the second World War. It also offers a fascinating insight into why that attempt failed, and the reasons why it has been so hard to see social actors, including women, in the post-World War inscriptions of technology. Despite their Mumfordian idealism – which priviledged culture and people in describing technological development – they built upon a tradition which saw mechanical, civil, and mining engineering as the pivot of technological change, to the exclusion of more anthropological and sociological con-

Figuur 41   Ford package engineers listening in wrapped attention to Madeline Pajas explaining computer solutions to design problems. Photograph featured in an article under the headline: "Women Engineers: Cure for Your Engineer Shortage?" *Product Engineering* (April 1968). Courtesy of University of Delft Library, Delft, The Netherlands.

cepts of technology that had allowed such earlier thinkers as Morgan, Gage, Mason, Smith, and Tarbell to perceive women in the technological project. The *Technology* entry emphasized capital-intensive industries and engineering, relegating the textile industry, where women and children worked in large numbers, to a minor plane, omitting household technologies altogether, and privileging the West, the U.S., and the twentieth century. And as the authors positioned mechanical and civil engineering at the center of their story, they traced the genealogy of these fields to ancient times. Strictly speaking, the causality and tradition inferred prescribed more than described these links. They implied that the history of engineering provided a road map for technological change. These notions were both exclusionary and limiting.

Emblematic, its lemma showed how the concept and definition of technology had undergone a dramatic transformation. After a century-long contest over their meanings, patents and machinery occupied center stage and stood at the heart of sexual and racial differentiation and class distinctions. Engineers were cast as its sole bearers at the expense of workers, African-Americans, and women.

Figure 42.   Photo in advertising style appearing in *Product Engineering* reporting on women in engineering in April 1968. Invisible but through her red-painted lips and nails, the woman engineer is left outside the frame and cast in opposition to male technical expertise represented by engineering knowledge and tools. Courtesy of University of Delft Library, Delft, The Netherlands.

The ink of the entry had barely dried when a younger generation of women, inspired by a new wave of feminism in the 1970s, questioned Ferguson and Kranzberg's carefully crafted genealogy of the history of technology. Published in 1979 and 1980, two collections entitled *Virgin and the Dynamo Revisited* and *Dea Ex Machina* challenged the engineering genesis of the field, arguing that its male domination excluded those historical actors operating outside its definitions where women must remain bystanders of a male-created stage. They linked back to some of the early critiques and struggles in the tradition from Mathilda Gage to Mary Beard; they questioned the banishments of corsets to the basements of the modern classification systems of technology; challenged the idea that women when they do appear on the scene as engineers and inventors function like *deae ex machina* as if appearing to come from nowhere. [Figure 42] In this male-constructed stage, women who enter the male-defined technical stage must always look like amateurs, or as Jan Zimmerman wryly remarked in those years, technology is what women don't do.[16] [Figure 41] But as Gage, Smith, Tarbell, and Beard had shown before, technology is a narrative production of our own times.

Notes

Introduction

1   John Markoff, "Computing in America: A Masculine Mystique," *The New York Times* (February 13, 1989): A1, 1-2; B10, 1-6; Michael Wines, "A Youth's Passion for Computers, Gone Sour," *The New York Times* (November 11, 1988): A1, 1; A28, 1, 28. Within one week, at least twelve articles in *The New York Times* were given journalistic center stage.

2   Alva T. Matthews, "Emily W. Roebling, One of the Builders of the Bridge," in *Bridge to the Future: A Centennial Celebration of the Brooklyn Bridge*, eds. Margaret Latimer, Brooke Hindle, and Melvin Kranzberg *Annals of the New York Academy of Sciences* (1984), 63-70; "Air Engineer," *The Christian Science Monitor* (February 25, 1943): 2; "Jersey Girl Has Job as Shipyard Engineer," *The New York Times* (May 8, 1949): 29; "Latham Girl Goes to New CE Position," *The Schenectady Gazette* (November 1, 1960); "Women Have Part in the History of General Electric," *General Electric News* (December 2, 1927); "Eyeball Engineer Draws 2d Glance," *The Philadelphia Inquirer* (March 30, 1970): 2. See also Clipping file, Ethel Lou Bailey, Hall of History Foundation, Schenectady, NY; Clipping file, Women Engineers, Schlesinger Library, Radcliffe College, Cambridge, MA.

3   For the locus classicus on the use of gender in historical studies, see: Joan W. Scott, "Gender: A Useful Category of Historical Analysis," *Journal of American History* (1986): 1053-75. Sandra Harding articualted a similar vocabulary in the field of feminist philosophy of science in Sandra Harding, *The Science Question in Feminism* (Ithaca: Cornell University Press, 1986), but see for the first use of the notion of gender in women's studies, Ann Oakley, *Sex, Gender and Society* (London: Temple Smith, 1972). For a discussion on the history of gendering of men for example, see: Elizabeth H. Pleck and Joseph H. Pleck, eds., *The American Man* (Englewood Cliffs, NJ: Prentice-Hall, 1980), Introduction; Harry Brod, ed., *The Making of Masculinities* (Boston: Allen and Urwin, 1987); Mark C. Carnes and Clyde Griffen, *Meanings for Manhood: Constructions of Masculinity in Victorian America* (Chicago: The University of Chicago Press, 1990); David D. Gilmore, *Manhood in the Making: Cultural Concepts of Masculinity* (New Haven: Yale University Press, 1990); Mark C. Carnes, *Secret Ritual and Manhood in Victorian America* (New Haven: Yale University Press, 1989); Gail Bederman, *Manliness & Civilization: A Cultural History of Gender and Race in the United States, 1880-1917* (Chicago: The University of Chicago Press, 1995).

4 Peter Meiksins, "Engineers in the United States: A House Divided," in *Engineering Labour: Technical Workers in a Comparative Perspective*, eds. Peter Meiksins and Chris Smith (London: Verso, 1996), 61-97; Terry S. Reynolds, ed., *The Engineer in America: A Historical Anthology from Technology and Culture* (Chicago: Chicago University Press, 1991); Edwin T. Layton Jr., *The Revolt of the Engineers: Social Responsibility and the American Engineering Profession* (Baltimore: The Johns Hopkins University Press, 1971), 3; David F. Noble, *America by Design: Science, Technology and the Rise of Corporate Capitalism* (New York: Cambridge University Press, 1977), 39; David M. Blank and George J. Stigler, *The Demand and Supply of Scientific Personnel* (New York: National Bureau of Economic Research, 1957), 4-5, 147; Alfred D. Chandler Jr., *The Visible Hand: The Managerial Revolution in American Business* (Cambridge: Harvard University Press, 1977).

5 U.S. Department of Labor, Woman's Bureau, *The Outlook for Women in Architecture and Engineering*, Bulletin 223 no. 5 (Washington, DC: Government Printing Office, 1948), 12; U.S. Department of Labor, Woman's Bureau, "Employment and Characteristics of Women Engineers," *Monthly Labor Review* (May 1956): 551-4; Edna May Turner, "Education of Women for Engineering in the United States, 1885-1952," (Ph.D. diss., New York University, 1954), 60, n. 1; File "Earliest efforts to organize 1929-1940," Box 146, Society of Women Engineers Collection, Walter Reuther Library, Wayne State University, Detroit, MI (SWE collection, hereafter); Marta Navia Kindya, *Four Decades of the Society of Women Engineers*, ed. Cynthia Knox Lang (Society of Women Engineers, n.d.); Martha Moore Trescott's, "A Progress Report," Manuscript, MIT, Institute Archives, Cambridge, MA, is based on women graduates in engineering at MIT and University of Illinois, Urbana, IL.

6 Letter Carnegie Institute of Technology to Hilda Counts, February 5, 1919, Box 146, SWE Collection; Juliet K. Coyle, "Evolution of a Species – Engineering Aide," *U.S. Woman Engineer* (April 1984): 25-26; B.K. Krenzer, "Early Leader still Has Impact: Lois Graham," *U.S. Woman Engineer* (July-August 1987); Clipping file, Women at GE, Hall of History Foundation, Schenectady, NY; Margaret W. Rossiter, *Women Scientists in America. Before Affirmative Action, 1940-1972* (Baltimore: The Johns Hopkins University Press, 1995), 267; "New Members of The Women's Engineering Society," *Woman Engineering Society* (1949), 315; "Emma Barth, PE, One of the First," *Woman Engineer* (March-April 1986): 17; U.S. Bureau of Census, *Abstract to the Eleventh Census* (Washington: U.S. Printing Office, 1890), 29; U.S. Bureau of the Census, *Twenty Censuses: Population and Housing Questions, 1790-1980* (Washington, DC: U.S. Printing Office, 1979), 29.

7 The term is Martha Banta's in her imaginative, *Taylored Lives: Narrative Productions in the Age of Taylor, Veblen and Ford* (Chicago: University of Chicago Press, 1993).

8 "A son (?) of the M.I.T.," Words from *Technique* (1897) to the melody of "The son of a Gambolier," arranged by F. F. Bullard, in *Techsongs: The M.I.T. Kommers Book*, rev. by a committee of the class of 1907 (Boston: O. Dibson Co., 1907 [1903]).

9 For example, see John Kasson, *Civilizing the Machine: Technology and Republican Values in America, 1776-1900* (New York: Penguin, 1977 [1976]). Many scholars have attempted to define the scope of technology without much success. See for sociology, Diana Crane, "In Search of Technology: Definitions of Technology in Different Fields –

An Essay Review" (paper prepared for presentation at the Mellon Foundation Seminar on Technology and Culture, February 4, 1987); for anthropology, Bryan Pfaffenberger, "Fetishised Objects and Humanised Nature: Towards an Anthropology of Technology," *MAN* 23 (June 1988): 236-52; for the history of technology, the work of John M. Staudenmaier, esp. *Technology's Storytellers: Reweaving the Human Fabric* (Cambridge: MIT Press, 1985); for economics, Nathan Rosenberg, *Perspectives of Technology* (New York: Cambridge University Press, 1976) and *Inside the Black Box: Technology and Economics* (New York: Cambridge University Press, 1982); See also, Michael Fores, "Technical Change and the 'Technology' Myth," *The Scandinavian Economic History Review* 30, 3 (1982): 167-88. Political scientist Langdon Winner forms an exception. Rather than decrying the absence of the term in early nineteenth-century discourse, he uses it as his point of departure in *Autonomous Technology: Technics-out-of-Control as Theme in Political Thought* (Cambridge: MIT Press, 1989 [1977]), 8-12.

10 Nina E. Lerman, "The Uses of Useful Knowledge: Science, Technology, and Social Boundaries in an Industrializing City," eds. Sally Kolhstedt and Ellen Longino, *Osiris. Special Issue Women, Gender, and the Science Question* (1997); Ronald Kline, "Construing 'Technology' as 'Applied Science'. Public Rhetoric of Scientists and Engineers in the United States, 1880-1945," *ISIS* 86 (1995): 194-221.

11 On the notion of keywords, see: Raymond Williams, *Keywords. A Vocabulary of Culture and Society*, revised edition (New York: Oxford University Press, 1976) [1953] and his earlier important book, *Culture and Society, 1780-1950* (New York: Harper and Row, 1983 [1958]). See also Daniel T. Rodgers, *Contested Truths. Keywords in American Politics since Independence* (New York: Basic Books, 1987). The literature on metaphors is vast, but see: Max Black, "More about Metaphor," and Thomas S. Kuhn, "Metaphor in Science," in *Metaphor and Thought*, ed. Andrew Ortony (Cambridge: Cambridge University Press, 1979), 19-43 and 409-19, respectively; George Lakoff and Mark Johnson, *Metaphors We Live By* (Chicago: University of Chicago, 1980); see also Nancy Leys Stepan's important contribution, "Race and Gender: The Role of Analogy in Science," *Isis* 77 (1986): 261-77.

12 S.v., "Technology," *Encyclopaedia Britannica*, 15th ed. (Chicago: Encyclopedia Britannica, 1978). *The Index to Periodical Literature* does not list the category separately until the 1930s. The Library of Congress's *Subject Headings* (1914-1986) does not list "technology" as a separate entry, but only as a heading to point the reader to other categories such as bridge building, chemical engineering, etc. In the sixth edition of 1957 there is a remarkable expansion of the category. "Technology" did not show up either in computer searches through various databases or in encyclopedic dictionaries published before the twentieth century.

13 *Encyclopaedia Britannica* (1978), 452.

14 *Encyclopaedia Britannica* (Chicago: Encyclopaedia Brittanica, 1929), s.v., "Technical Education" and "Engineering Education." Cf. *Encyclopaedia Americana* (1920), s.v., "Schools of Technology."

Chapter 1

1   Based on various database searches and a survey of various editions of major dictionaries and encyclopedias including John Kersey's *Dictionarium Anglo-Britannicum* (London: Phillips, 1708); *Cyclopaedia* (London: Longman, 1819); John Pickering, *A Vocabulary or Collection of Words and Phrases Which Have Been Supposed to be Peculiar to the United States etc.* (Boston: Cummings and Hilliard, 1816); Noah Webster's *American Dictionary of the English Language* (New York: Converse, 1828 examined through 1972); *Century's Dictionary and Cyclopaedia* (1911); *The Encyclopaedia Britannica*, first through fourteenth editions; *New Americanized Encyclopaedia Britannica* (1896-1904); *The New Encyclopaedia Britannica* (Chicago: Encyclopaedia Britannica, 1974-1985); and *The Oxford English Dictionary* (Oxford: Clarendon Press, 1933). Significantly, Raymond Williams in *Keywords. A Vocabulary of Culture and Society* (New York: Oxford University Press, 1976), did not include technology as a keyword in either this or a later edition of his book.

2   Bruce Sinclair, *Philadelphia's Philosopher Mechanics: A History of the Franklin Institute, 1824-1865* (Baltimore: The Johns Hopkins University Press, 1974); Daniel T. Rodgers, *Contested Truths. Keywords in American Politics Since Independence* (New York, Basic Books, 1987), 17, Chapter 1. For the discussion on both sides of the Atlantic, see also Frank E. Manuel, *The Prophets of Paris* (Cambridge: Harvard University Press, 1962), Chapter 3; Peter Gay, *The Enlightenment: An Interpretation* (London: Wildwood House, 1973 [1966]), I, 178-83, and II, 252-3; Franklin L. Baumer, *Modern European Thought: Continuity and Change in Ideas, 1600-1950* (New York: Macmillan, 1977); Robert K. Merton, *Science, Technology and Society in Seventeenth-Century England* (New York: Harper and Row, 1970), 18-22.

3   See Jeanne Boydston's critique of this limited understanding of women's economic role in "'To Earn her Daily Bread': Housework and Antebellum Working-Class Subsistence," *Radical History Review* 35 (April 1986): 7-25, and her *Home and Work: Housework, Wages and the Ideology of Labor in the Early Republic* (New York: Oxford University Press, 1990).

4   Edith Abbott, *Women in Industry: A Study in American Economic History* (New York: D. Appleton, 1910), Chapter 4; Paula Baker, "The Domestication of Politics: Women and American Political Society, 1780-1920," *American Historical Review* 89 (June 1984): 620-47 and *The Moral Framework of Public Life: Gender, Politics and the State in Rural New York, 1870-1920* (New York: Oxford University Press, 1991), 24-55; Judith A. McGaw, *Most Wonderful Machine: Mechanization and Social Change in Berkshire Paper Making, 1801-1885* (Princeton: Princeton University Press, 1987); John Kasson, *Civilizing the Machine: Technology and Republican Values in America* (New York: Grossman, 1976), 27. Cf. Frances Gouda, *Poverty and Political Culture: The Rhetoric of Social Welfare Netherlands and France, 1815-1854* (Lanham, MD: Roman Littlefield, 1995), 148.

5   David F. Noble, *A World Without Women: The Christian Clerical Culture of Western Science* (New York: Alfred A. Knopf, 1992), Chapter 10.

6   Tenson Coxe, "An Address to an Assembly of the Friends of American Manufacturers, Convened for the Purpose of Establishing a Society for the Encouragement of Manu-

facturers and the Useful Arts, Read in the University of Pennsylvania, on Thursday the 9th of August 1787," (Philadelphia, 1787) in *The Philosophy of Manufacturers: Early Debates over Industrialization in the United States*, eds. Michael Brewster Folsom and Steven D. Lubar, *Documents in American Industrial History* I (Cambridge, MA: MIT Press, 1982), 42; see also Benjamin Rush, "A Speech Delivered in Carpenter's Hall, March 16th, Before the Subscribers Towards a Fund for Establishing Manufactories of Woolen, Cotton, and Linnen, in the City of Philadelphia. Published at the Request of the Company," *The Pennsylvania Evening Post* (April 11, 1775), 133, and (April 13, 1775), 137-8; Alexander Hamilton, *Report of the Secretary of the Treasury of the United States, on the Subject of Manufactures. Presented to the House of Representatives, December 5, 1791* (Philadelphia, 1791).

7   Sinclair, *Philadelphia's Philosopher Mechanics*; Nina E. Lerman, "The Uses of Useful Knowledge: Science, Technology, and Social Boundaries in an Industrializing City," *Osiris. Special Issue Women, Gender, and the Science Question*, eds. Sally Kolhstedt and Ellen Longino (1997). Ronald Kline, "Construing 'Technology' as 'Applied Science'. Public Rhetoric of Scientists and Engineers in the United States, 1880-1945," *ISIS* 86 (1995): 194-221; David Hounshell, "Edison and the Pure Science Ideal in America," *Science* 207 (February 8, 1980): 612-7. Arthur P. Molella and Nathan Reingold, "Theorists and Ingenious Mechanics: Joseph Henry Defines Science," *Science Studies* 3 (1973): 322-35. In Britain, scientists enlisted manufacturers and engineers as rank-and-file members to show science's material alliance to the idea of progress, but they were careful to keep them otherwise at bay. Jack Morrell and Arnold Thackray, *Gentlemen of Science. Early Years of the British Association for the Advancement of Science* (London: Clarendon Press, 1981), 256-66.

8   Bruce Sinclair, "Inventing a Genteel Tradition: MIT Crosses the River," in *New perspectives on Technology and American Culture*, ed. Bruce Sinclair, American Philosophical Society Library Publication 12 (Philadelphia: American Philosophical Society, 1986), 1-18; Kline, "Construing 'Technology' as 'Applied Science'." For an indication, in chronological order, of technology as a label of self-identification: Massachusetts Institute of Technology (1861); Stevens Institute of Technology (1870); Georgia Institute of Technology (1885); Clarkson College of Technology (1896); Carnegie-Mellon (1912); California Institute of Technology (1920); Lawrence Institute of Technology (1932). Most other institutions were either established or renamed after the second World War including such schools as: Rochester Institute of Technology; Florida Institute of Technology; Illinois Institute of Technology, see *American Universities and Colleges* (New York: Walter de Gruyter, 1983 [12th edition]).

9   Edward Hazen, *Popular Technology: or, Professions and Trades* (New York: Harper and Brothers, 1841), Introduction. A similar and broad usage is exemplified in an editorial, "The Technology of Ceramic Art," *The New Century for Woman* 23 (Saturday, October 24, 1876): 179 and in a paper by P. Geddes, "Economics and Statistics, Viewed from the Standpoint of the Preliminary Sciences," *Nature* 24 (May-October, 1881): 523-6.

10 I have not been able to verify Bigelow's resurrection of the term from old dictionaries, but see the claim made by Jennifer Clark, "The American Image of Technology from the Revolution to 1840," *American Quarterly* 39, 3 (Fall 1987): 431-49.

11 Bruce Seeley, "Research, Engineering, and Science in American Engineering Colleges, 1900-1960," *Technology and Culture* 34 (1993): 344-86; Mark H. Rose, "Science as an Idiom in the Domain of Technology," *Science and Technology Studies* 5, 1 (1987): 3-11.

12 I am grateful to Nina Lerman for this insight, personal communication, June 1998.

13 Booker T. Washington, "Industrial Education," *Annals of American Academy of Political and Social Science* 33, 1 (1909) and *Up from Slavery* (New York: A.L. Burt, 1901); W.E.B. Du Bois, *The Souls of Black Folk* (1903).

14 Robert W. Rydell, "The Literature of International Expositions," in *The Books of the Fairs. Material about the World's Fairs, 1834-1916, in the Smithsonian Institution Libraries* (Chicago: American Library Association, 1992), 1-62; Eugene S. Ferguson, "Expositions of Technology, 1851-1900," in *Technology in Western Civilization*, eds. Melvin Kranzberg and Carroll Pursell, Jr. (New York: Oxford University Press, 1967), 706-26 and "Power and Influence: The Corliss Engine in the Centennial Era," in *Bridge to the Future: A Centennial Celebration of the Brooklyn Bridge*, eds. Margaret Latimer et al., *Annals of the New York Academy of Sciences* 424 (1984), 225-46; Robert C. Post, "Reflections of American Science and Technology at the New York Crystal Palace Exhibition of 1853," *Journal of American Studies* 17 (1983): 337-56.

15 Robert W. Rydell, *All the World's a Fair: Visions of Empire at American International Expositions, 1876-1916* (Chicago: The University of Chicago Press, 1984), 2; Rydell, "The Literature of International Expositions," 42-3; "The Fan Dance of Science; American World's Fairs in the Great Depression," *Isis* 76, 284 (1985): 525-42.

16 Lewis H. Morgan, *Ancient Society: or, Researches in the Lines of Human Progress from Savagery through Barbarism to Civilization* (New York: H. Holt, 1877), 39-40, and chapters 1-3; Robert E. Bieder, *Science Encounters the Indian* (Norman: The University of Oklahoma Press, 1986), 194-246. On the basis of Marx's readings and notes of Morgan's book, Friedrich Engels (1820-1895) was able to write *The Origin of the Family, Private Property in the State in Light of the Researches of Lewis H. Morgan.*

17 Frederick Beach, "Patents," *Encyclopedia Americana*; *DAB*, s.v., "Frederick Converse Beach"; Eugene S. Ferguson, "Technical Journals and the History of Technology," in *In Context: History and the History of Technology. Essays in Honor of Melville Kranzberg*, eds. Stephen H. Cutcliffe and Robert C. Post (Bethlehem: Lehigh University Press, 1989), 53-70. See also Michael Adas, *Machines as the Measure of Men: Science, Technology, and Ideologies of Western Dominance* (Ithaca: Cornell University Press, 1989), 312-13.

18 The crank metaphor feminists appropriated came from Henry Thoreau, "Paradise (to be) regained," *United States Magazine, and Democratic Review* 13, 65 (November 1843) which critically reviewed J. A. Etzler, *Paradise Within the Reach of All Men* published in 1833.

19 For a treatment on the collaborations and tensions between feminists and women inventors, see Jeanne Madeline Weimann, *The Fair Women: The Story of the Woman's Building, World's Columbian Exposition Chicago 1893* (Chicago: Academy Chicago, 1981), 429-32; Anne L. Macdonald, *Feminine Ingenuity: Women and Invention in America* (New York: Ballantine Books, 1992), 103-16; Deborah J. Warner, "Women Inventors at the Centennial," in *Dynamos and Virgins Revisited*, ed. Martha Moore Trescott (Metuchen, NJ: Scarecrow Press, 1979), 102-19. For a exhaustive listing of women in-

ventors in nineteenth- and twentieth-century America, see: Autumn Stanley, *Mothers and Daughters of Invention: Notes for a Revised History of Technology* (Methuen, NJ: Scarecrow Press, 1993).

20 M. E. Joslyn Gage, "Woman as inventor," issued under the auspices of the New York State Woman Suffrage Association, in *Woman Suffrage Tracts* 1 (1870): 7. The history of Greene's invention can be traced in the papers of Eli Whitney, Sterling Memorial Library, Manuscripts and Archives, Yale University, New Haven, CT. Two years prior to the pamphlet, Gage had already argued for Greene's contribution to the cotton gin in the *Revolution*, the National Woman Suffrage Association; *Notable American Women, 1607-1950* (Cambridge: Harvard University Press, 1971), s.v., "Gage."

21 Matilda Joslyn Gage, "Woman as inventor," *North American Review* 136 (May 1883): 478-89.

22 Rydell, *All the World's Fair*, 14-5. See also Adas, *Machines as the Measure of Men*; Bryan Pfaffenberger, "Fetishised Objects and Humanised Nature: Toward an Anthropology of Technology," *Man* 23 (June 1988): 236-52.

23 As cited in Warner, "Women Inventors at the Centennial," 102; Virginia Grant Darney, "Women and World's Fairs: American International Expositons, 1876-1904," (Ph.D. diss., Emory University, 1982).

24 *The New Century for Women* 1 (Philadelphia, 1876), 151; United States Centennial Commission, *International Exhibition, 1876. Reports and Awards*, ed. Francis A. Walker (Washington: Government Printing Office, 1880). *Tribune Guide to the Exhibition* (New York: The Tribune, 1876), 44; Darney, "Women and World's Fairs"; Elizabeth Cady Stanton, Susan B. Anthony, and Matilda Joslyn Gage, eds., *History of Woman Suffrage* Vol. 3, 1876-1885 (Rochester: Charles Mann, 1886), 54-5. Robert C. Post, ed., *1876: A Centennial Exhibition Catalogue to Exhibition* (Washington DC: Smithsonian Institution, 1976), 167.

25 Darney, "Women and World's Fairs"; Deborah Warner, "The Woman's Pavilion," in Post, *1876: A Centennial Exhibition*, 171.

26 Stanton et al., *History of Woman Suffrage*, 54.

27 Gage, "Woman as Inventor" (1880), 488.

28 Ida M. Tarbell, "Women as Inventors," *Chautauquan* 7, 6 (March 1887): 355-7; *Dictionary of American Women Historians* (Westport, CT: Greenwood Press, 1996) and David M. Chalmer *Notable American Women* (Westport, CT: Greenwood Press, 1996), s.v., "Tarbell". See also the discussion in Macdonald, *Feminine Ingenuity*, 118-9 and her (incorrect) dating of Tarbell's article.

29 Only one issue of the magazine, *The Woman Inventor*, was published. See for Smith's acknowledgement of Gage: editorial of 1, 1 (April 1890): 2. Little is known about Charlotte Smith, but see *The New York Times* Obituary (December 6, 1917): 13; 4. See also Weimann, *The Fair Women*, 507-11; Carroll Pursell, "Cover Design: Women Inventors in America," *Technology and Culture* 22 (1981): 545-48, 547, fn 11; U.S. Patent Office, *Women Inventors to Whom Patents Have Been Granted by the United States, 1790 to July 1, 1888, with Appendices to March 1, 1895* (Washington, DC: Government Printing Office, 1888, 1895).

30 Autumn Stanley, "The Patent Office Clerk as Conjurer: The Vanishing Lady Trick in a Nineteenth-century Historical Source," in *Women, Work, and Technology: Transforma-*

Making Technology Masculine

tions, ed. Barbara Wright Drygulski and et al. (Ann Arbor: The University of Michigan, 1987), 118-36, 128.
31 Editorial, *Woman Inventor*, 1.
32 The famous Boas-Mason exchange on inventions: Franz Boas, "The Occurrence of Similar Inventions in Areas Widely Apart," *Science* 7 (1887): 485-6 and Otis T. Mason, "The Occurrence of Similar Inventions in Areas Widely Apart," *Science* 9 (1887): 534-5. George W. Stocking considers this debate "fundamental to Boas' anthropology," in *A Franz Boas Reader: The Shaping of American Anthropology, 1883-1911* (Chicago: The University of Chicago Press, 1973), 1-20.
33 Otis Tufton Mason, "The Birth of Invention," in *Patent Centennial Celebration, Proceedings and Addresses. Celebration of the Beginning of the Second Century of the American Patent System at Washington City, DC, April 8, 9, 10, 1891* (Washington, DC: Press of Gedney & Roberts, 1892), 403-12.
34 On the expositions, see: Rydell, *All the World's a Fair*, 23-4; 57-60; 97-100.
35 Otis Tufton Mason, *Woman's Share in Primitive Culture* (New York: D. Appleton, 1894), 11. For an excerpt see, "Woman as an Inventor and Manufacturer," *Popular Science Monthly* 47 (May 1895): 92-103. See also, Louise Michele Newman, ed., *Men's Ideas/Women's Realities: Popular Science, 1870-1915,* (New York: Pergamon, 1985), foreword by Ruth Hubbard.
36 Mason, *Woman's Share in Primitive Culture*, 2; Nancy Leys Stepan, "Race and Gender: The Role of Analogy in Science," *Isis* 77 (1986): 261-77.
37 Jeanne Madeline Weimann, "The Great 1893 Woman's Building: Can We Measure Up in 1992?" MS Magazine 41 (March 1983): 65-7 and *The Fair Women*, 429-33; Macdonald, *Feminine Ingenuity*.
38 Ann Massa, "Black Women in the 'White City'," *Journal of American Studies* 8 (1974): 319-37; Eliott M. Rudwick and August Meier, "Black Man in the 'White City': Negroes and the Columbian Exposition, 1893," *Phylon* 26 (Winter 1965): 354-61; Gail Bederman, *Manliness & Civilization: A Cultural History of Gender and Race in the United States, 1880-1917* (Chicago, The University of Chicago Press, 1995), 31-41.
39 J.B. Bury, *The Idea of Progress: An Inquiry into Its Origin and Growth* (New York: Dover Publications, 1932 [1927]), intro. by Charles A. Beard, xxii.
40 Henry L. Mencken, "Professor Veblen and the Cow," *Smart Set* 59 (May 1919): 138-44; Lewis A. Coser, *Masters of Sociological Thought: Ideas in Historical and Social Context* (New York: Harcourt Brace Jovanovich, 1977 [2nd ed.]), 263-302.
41 For an indication of his influence, see Coser, *Masters of Sociological Thought*; Lewis Mumford, *Technics and Civilization* (New York: Harcourt, Brace, Janovavich, 1934), 25, 55, 96, 226, 284, 317, 354, 366, 401, 472, 475. Stuart Chase wrote in his introduction to *Theory of the Leisure Class* (New York: Modern Library, 1934 [1899]): "Thorstein Veblen was one of my idols as a young man. When a new book of his was published I secured it at once and read it many times."
42 John Dos Passos, *The Big Money* (New York: Washington Square Press, 1961 [1930]), 107.
43 Mencken, "Professor Veblen and the Cow." See also a reviewer's comment on Veblen's peculiar style: "Is it too much to hope that some of his followers will translate this book into English readable to the economic layman?" Anon., "Mr. Veblen's Economics,"

Springfield Republican 18 (March 1920): 539. See also Stephen S. Conroy, "Thorstein Veblen's Prose," *American Quarterly* 20 (Fall 1968): 605-15.

44 Veblen, *The Theory of the Leisure Class*, 149, and 141-2, 138, 172, 182-3.

45 See also John P. Diggins, *The Bard of Savagery: Thorstein Veblen and Modern Social Theory* (New York: The Seabury Press, 1978): Chapter 8; Daniel Bell, "Veblen and the Technocrats: On *The Engineers and the Pricesystem*," in *The Winding Passage* (New York: Basic Books, 1980 [1963]): 69-90, p. 88, fn. 32. See also Edwin T. Layton Jr., "Veblen and the Engineers," *American Quarterly* 14, 1 (Spring 1962): 64-72.

46 Thorstein Veblen, "Using the I.W.W. to Harvest Grain, Unpublished Paper (1918)," introd. by Joseph Dorfman, *Journal of Political Economy* (December 1932) and "An Unpublished Paper on the I.W.W. by Thorstein Veblen," introd. by Joseph Dorfman *Journal of Political Economy* (December, 1943).

47 "The Industrial Encyclopedia," *The One Big Union Monthly* 1, 12 (December 1919): 15. For more of the episode, see the articles appearing in *The One Big Union Monthly*: A.G., "Are You Prepared to Manage Industry?" 1, 5 (May 1919): 42; John Sandgren, "The I.W.W. Needs an Industrial Encyclopedia," 1, 11 (November 1919): 42-4; "Agricultural Workers," 1,12 (December 1919): 53; "The Twelfth Annual Convention of the I.W.W.," 2,5 (May 1920): 5; "Agriculture. The World's Basic Industry and Its Workers," 6 (1920): 61; R. Bruner, "Get More Technical Knowledge," 2, 7 (June 1920): 48; Ralph Chaplin, "The Bureau of Industrial Research and Its Work," 2, 8 (July 1920): 56-7; "The 'Harvest Stiff' of Ancient Days: A Chapter of Suppressed History," 2, 8 (August 1920): 17; Howard Scott, "The Scourge of Politics in a Land of Manna," 2, 9 (September, 1920): 14-6; [Howard Scott], "Political Schemes in Industry by an Industrial Engineer," 2,10 (October 1920): 6-10; "Special Meeting of the General Executive Board," 2, 11 (November, 1920): 62. While I.W.W. leadership had read Veblen's article "Soviet of Engineers," *The Dial*, they also seemed to have reached similar conclusions independently of Veblen. Ralph Chaplin, *Wobbly: The Rough-and-Tumble Story of an American Radical* (New York: Da Capo 1972 [1948]), 295; Robert L. Tyler, "The I.W.W. and the Brainworkers," *American Quarterly* 15 (Spring 1963): 41-51, pp. 44-5.

48 Thorstein Veblen, *The Engineers and the Pricesystem* (New York: B. W. Huebsch, 1921), 86-9.

49 Veblen's intellectual debt to anthropologists has not been discussed extensively by Veblen scholars, but emerges quite clearly from *The Theory of the Leisure Class*. See also: Bernard Rosenberg, "A Clarification of Some Veblenian Concepts," *American Journal of Economics and Sociology* 12 (January 1953): 179-87; on the influence of Spencer and Sumner, and Veblen's reworkings, see Richard Hofstadter, *Social Darwinism in American Thought, 1869-1915* (Boston: Beacon Press, rev. 1955 [1944]), 152-6, 168-9.

50 Veblen, *Engineers*, 52.

51 For an indication of his use of these words, see: 24, 28, 30, 33, 40-3, 52, 59-61, and 69. For an earlier but sparse use, see Thorstein Veblen, "The Place of Science in Modern Civilization," *The American Journal of Sociology* 11 (March 1906).

52 Kline, "Construing 'Technology' as 'Applied Science'."

53 Unless otherwise indicated, the following paragraph is based on Amy Sue Bix, "Inventing Ourselves Out of Jobs: America's Depression Era Debate Over Technological Unemployment" (Ph.D. diss., Johns Hopkins University, 1994), esp. Chapters 4-5.

54 William E. Akin, *Technocracy and the American dream: The technocracy movement, 1900-1941* (Berkeley: University of California Press, 1977).

55 Warren I. Susman, "The People's Fair: Cultural Contradictions of a Consumer Society," in *Culture as History: The Transformation of American Society in the Twentieth Century* (New York: Pantheon, 1984): 211-29; Robert W. Rydell, *World of Fairs: The Century-of-Progress Expositions* (Chicago: University of Chicago Press, 1993); Bix, "Inventing Ourselves."

56 Karl T. Compton, "Technology's Answer to Technocracy,' in *For and Against Technocracy: A Symposium*, ed. J. George Frederick (New York: Business Bourse, 1933), 77-93, p. 77.

57 Bix, "Inventing Ourselves," 471-2.

58 Arthur D. Little, "Technocracy vs. Technology," *Commercial and Financial Chronicle* 136 (January 21, 1933): 435-7. Little and Millikan responded to governmental researchers analyses of technological unemployment discussed in the *Monthly Labor Review* for example, U.S. Bureau of Labor Statistics, United States Bureau of Labor Statistics, "Effect of Technological Changes upon Employment in the Amusement Industry," 32, 2 (August, 1931): 261-7 and "Effects of Technological Changes Upon Employment in the Motion-Picture Theaters of Washington, DC," 33, 3 (November 1931): 1005-18; Alvin Hansen, "Institutional Frictions and Technological Unemployment," *Quarterly Journal of Economics* 45 (August 1930-31): 684-97; Emil Lederer in the first edition of the *Encyclopaedia of the Social Sciences*, eds. Edwin R. A. Seligman and Alvin Johnson (New York: Macmillan Company, 1934) included the first and only essay on technology and discussed the issue of technological unemployment. Significantly, the Encyclopaedia, whose contributors had many personal and intellectual ties to Veblen and the technocracy movement, sprang up around the community of the New School in New York. See also Benjamin M. Anderson, "Technological Progress, The Stability of Business and the Interests of Labor," *The Chase Economic Bulletin* (issued by the Chase National Bank of the City of New York) 17, 2 (13 April 1937): 3-35.

59 See Arthur D. Little, "'The Fifth Estate,'" (address delivered at the Franklin Institute Centenary) *The Atlantic Monthly* 134 (December 1924): 771-81.

60 Willliam F. Ogburn, *Living with Machines* (Chicago: American Library Assocation, 1933); Abbott Payson Usher, *A History of Mechanical Inventions* (New York: McGraw-Hill, 1929); William Ogburn with S.C. Gilfillan, "The Influence of Invention and Discovery," in *Recent Social Trends in the United States* (New York: McGraw-Hill, 1933); S.C. Gilfillan, *The Sociology of Invention* (Cambridge: MIT Press, 1935) and "Social Effects of Inventions," in *Technological Trends and National Policy Including the Social Implications of New Inventions*, National Resource Committee (Washington, DC: U.S. Government Printing Office, 1937), 24-66. The change from machine discourse to technology can also be traced through Ogburn's work from *Living with Machines* of 1933 to his "National Policy and Technology," in National Resources Committee, *Technological Trends and National Policy Including the Social Implications of New Inventions* (Washington, DC: U.S. Government Printing Office, 1937), 3-14; Arthur P. Molella, "The First Generation: Usher, Mumford, and Giedion," Cutliffe and Post, *In Context*, 88-105.

61 See Boas-Mason debate on the 'Occurence of Similar Inventions' in Stocking, *Boas Reader*; David McGee, "Making Up Mind: The Early Sociology of Invention," *Technology and Culture* 36, 4 (October 1995): 773-801.

62 Rydell, *World of Fairs*, chapter 6; Rydell, "The Literature of International Expositions," 42-3; Susman, "The People's Fair."

CHAPTER 2

1   J.A.L. Waddell, "The Advisability of Instructing Engineering Students in the History of the Engineering Profession," *Proceedings of the Society for the Promotion of Engineering Education* 11 (1903): 193-217.

2   The literature on engineering is vast, but I have benefitted most from Peter Meiksins's work, see, for example: "Engineers in the United States: A House Divided," in *Engineering Labour: Technical Workers in a Comparative Perspective*, eds. Peter Meiksins and Chris Smith (London: Verso, 1996), 61-97.

3   On the social background of engineers, see: William E. Wickenden, *Report of the Investigation of Engineering Education* (Pittsburgh: Society for the Promotion of Engineering Education, 1930), I, 162; Carolyn Cummings Perrucci, "Engineering and the Class Structure," and Robert L. Eichhorn, "The Student Engineer," in *The Engineers and the Social System*, eds. Robert Perrucci and Joel Gerstl (New York: John Wiley and Sons, 1969). For engineer's status as a middle-class profession, see: Gallup, July 1, 1953, in *The Gallup Poll; Public Opinion 1935-1971* (New York: Random House, 1972), 1152, 1779.

4   "Organization for Engineers," *The Monad* 3, 2 (February 1918): 7-10, p. 8. A historically informed class analysis of engineers has been a much neglected area in the literature with the exception of Peter F. Meiksins's important work on the issue, see: "Scientific Management and Class Relations – A Dissenting View," *Theory and Society* 13, 2 (1984): 177-209. Stuart M. Blumin, "The Hypothesis of Middle-class Formation in Nineteenth-century America: A Critique and Some Proposals," *American Historical Review* (April 1985): 299-328; his *The Emergence of the Middle Class: Social Experience in the American City, 1760-1900* (Cambridge: Cambridge University Press, 1989); Olivier Zunz, *Making America Corporate, 1870-1920* (Chicago: The University of Chicago Press, 1990), and Daniel T. Rodgers, *The Work Ethic in Industrial America, 1850-1920* (Chicago: The University of Chicago Press, 1974), (esp. Chapter 1) are welcome additions to literature on the stratum of middle management to which most engineers belonged.

5   Edwin T. Layton Jr., *The Revolt of the Engineers: Social Responsibility and the American Engineering Profession, Baltimore* (Baltimore: The Johns Hopkins University Press: 1986[1971]); Peter F. Meiksins, "'The Revolt of the Engineers' Reconsidered," *Technology and Culture* 29, 2 (1988): 219-46; "Professionalism and Conflict: The Case of the American Association of Engineers," *Journal of Social History* 19, 3 (Spring 1986): 403-22 and "Professional Autonomy and Organizational Constraint – The Case of Engineers," *Sociological Quarterly* 30, 4 (1989): 561-85.

6   Alfred West Gilbert, *Colonel A. W. Gilbert. Citizen-Soldier of Cincinnati* (Cincinnati: Historical and Philosophical Society of Ohio, 1934), 32.

7 James Worrall, *Memoirs of Colonel James Worrall, Civil Engineer, with an Obituary Post-script by a Friend* (Harrisburg, PA: E.K. Meyer, 1887), 25.

8 For patterns of on-the-job training in corporations and the federal and military internal improvement projects see, William H. Goetzmann, *Army Exploration in the American West, 1803-1863* (New Haven: Yale University Press, 1959), 6-10; Thomas G. Manning, "The United States Geological Survey," in *Government Agencies*, ed. Donald R. Whitnach (Westport, CT: Greenwood Press, 1983), 548-53; Cedric E. Gregory, *A Concise History of Mining* (New York: Pergamon Press, 1980); Clarke C. Spence, *Mining Engineers & the American West: The Lace-Boot Brigade, 1849-1919* (New Haven: Yale University Press, 1970), 5, 60, 209, and Chapters 6, 9; see also, James M. Searles, *Life and Times of a Civil Engineer* (Cincinnati: privately printed, 1893), 17.

9 Peter Way, *Common Labor: Workers and the Digging of North American Canals, 1780-1860* (Baltimore: The Johns Hopkins University Press, 1993), 14.

10 Nina E. Lerman, "From 'Useful Knowledge' to 'Habits of Industry': Gender, Race and Class in Nineteenth Century Technical Education" (Ph.D. diss, University of Pennsylvania, 1993), Chapter 1 and "'Preparing for the Duties and Practical Business of Life': Technological Knowledge and Social Structure in Mid-19th-Century Philadelphia," *Technology and Culture* 38, 1 (January 1997): 31-59; Ian Quimby, *Apprenticeship in Colonial Philadelphia* (New York, 1985 [1963]); W. J. Rorabaugh, *The Craft Apprentice: from Franklin to the Machine Age in America* (Oxford: Oxford University Press, 1986). Classic apprenticeship was moribund by 1865, but its symbolic functions were revitalized in practical training patterns.

11 Robert Ridgway, *Robert Ridgway*, with Isabelle Law Ridgway (New York: privately printed, 1940), Chapters 1-7; and "My Days of Apprenticeship," *Civil Engineering* 8, 3 (September 1938): 601-4.

12 Philip Scranton, "Learning Manufacture: Education and Shop-floor Schooling in the Family Firm," *Technology and Culture* 27, 1 (January, 1986): 40-62; Spence, *Mining Engineers*, 23; Rorabaugh, *The Craft Apprentice*.

13 Raffe Emerson to Harrington Emerson, August 21, 1904, Eng. Z1, 1904, Box 6, file 1, Harrington Emerson Papers, PTA Historical Collection, Pennsylvania State University. I am very grateful to John Brown for sharing this extraordinary letter with me. John K. Brown, *The Baldwin Locomotive Works, 1831-1915: A Study of American Industrial Practice* (Baltimore: The Johns Hopkins University Press, 1995).

14 David Montgomery, "Workers' Control of Machine Production in the Nineteenth Century," in *Workers' Control in America* (New York: Cambridge University Press, 1979), 1-31.

15 Samuel Haber, *Efficiency and Uplift. Scientific Management in the Progressive Era, 1890-1920* (Chicago: Chicago Press, 1964), esp. Chapter 2; Monte Calvert, *The Mechanical Engineer in America, 1830-1910* (Baltimore: The Johns Hopkins University Press, 1967), 8; Frederick Winslow Taylor, *The Principles of Scientific Management* (New York: W. W. Norton [1911]), 26-7. Paul Willis, "Masculinity, the Wage Form, and Factory Labor" in *Working-class Culture*, eds. John Clarke Clarke et al. (London: Hutchinson, 1967), 185-98 suggests the connection. See also, Blumin, "Hypothesis of the Middle-Class Formation"; George M. Frederickson, *The Inner Civil War. Northern Intellectuals and the Crisis of the Union* (New York: Harper & Row, 1965); For engineers

with health problems between 1875 and 1880, see for example: Searles, *Life and Times*, and *DAB*, s.v., "Frederick W. Taylor", "Robert Thurston", and "J.A.L. Waddell."

16 Editorial, "The Title of the Engineer," *The American Machinist* 16 (1895).

17 As cited in Spence, *Mining Engineers*, 12; John W. Leonard, ed., *Who's Who in Engineering 1922-1923* (New York: John W. Leonard, 1922), s.v., "Rickard."

18 H. Pennington to Hamilton M. Barksdale, December 3, 1915 and December 3, 1916; also Mr. Robert E. Spragins to Mr. T. C. du Pont, September 1913; T. Coleman du Pont to Hamilton M. Barksdale, September 18, 1913; William M. Barksdale to T. Coleman du Pont, September 26, 1913; Hamilton M. Barksdale to William G. Ramsay, September 19, 1913, Box 1003A, File 6, DuPont Papers, Hagley Museum and Library, Wilmington, DE. On the reorganization at DuPont: Olivier Zunz, *Making America Corporate*, 70; JoAnne Yates, *Control Through Communication: The Rise of System in American Management* (Baltimore: Johns Hopkins University, 1989), 229-70. Cf. William LeRoy Emmet, *The Autobiography of an Engineer* (Albany: Fort Orange Press, 1931), 24.

19 Few scholars, if any, have looked at engineering manliness in a sustained manner, but see for some suggestive leads Scranton, "Learning Manufacture," on this process of shop-floor socialization, and Carroll Pursell, "The Long Summer of Boy Engineering," in *Possible Dreams. Enthusiasm for Technology in America*, ed. Johan L. Wright (Dearborn, MI: Henry Ford Museum and Greenfield Village, 1992): 35-43; and "The Construction of Masculinity and Technology," *Polhem* 11 (1993): 206-19.

20 O.H. Garman, "The Engineer's Place in Society," *The Monad* 2, 12 (December 1917): 8-9; Calvert, *The Mechanical Engineer*, 72-3; Scranton, "Learning Manufacture"; Terry S. Reynolds, "Defining Professional Boundaries: Chemical Engineering in the Early Twentieth Century," *Technology and Culture* 27, 4 (October, 1988): 694-716, 710-11.

21 See especially John Fritz, *The Autobiography of John Fritz* (New York: John Wiley & Sons, 1912), as discussed in greater detail in the next chapter.

22 Harvey Green, *Fit for America: Health, Fitness, Sport, and American Society* (Baltimore: The Johns Hopkins University Press, 1986); Gail Bederman, *Manliness & Civilization: A Cultural History of Gender and Race in the United States 1880-1917* (Chicago: The University of Chicago Press, 1995); Elizabeth H. Pleck and Joseph H. Pleck, eds., *The American Man* (Englewood Cliffs, NJ: Prentice Hall, 1980), Introduction; E. Anthony Rotundo, *American Manhood: Transformations in Masculinity from the Revolution to the Modern Era* (New York, Basic Books, 1993); James R. McGovern, "David Graham Phillips and the Virility Impulse of Progressives," *New England Quarterly* 39 (1966): 334-55.

23 *The Monad* 1, 8 (December 1916): 6. See also "Engineers Loyal in Strike," *Professional Engineer* 5, 6 (1920): 17; Meiksins, "Engineers in the United States"; Spence, *Mining Engineers*, 182, 192; Zunz, *Making America Corporate*, 61-4.

24 Franklin DeR. Furman, "Shall My Boy Become an Engineer?" *Scientific American* 106 (April 6, 1912): 314-5. For useful overview, see Peter Lundgreen, "Engineering Education in Europe and the U.S.A., 1750-1930: The Rise to Dominance of School Culture and the Engineering Professions," *Annals of Science* 47 (1990): 33-75.

25 David M. Blank and George J. Stigler, *The Demand and Supply of Scientific Personnel*, National Bureau of Economic Research, General Series, 62 (New York: National Bu-

reau of Economic Research, 1957), II, 89-90, Tables 5 and 35. See also Spence, *Mining Engineers*, 18-23, 52; Ernes Havermann and Patricia Slater West, eds., *They Went to College* (New York: Harcourt, 1952); Edwin T. Layton Jr., "Science, Business and the American Engineer," in *Engineers and the Social System*, 51-72, p. 53, fn. 4; David F. Noble, *America by Design: Science, Technology and the Rise of Corporate Capitalism* (New York: Cambridge University Press, 1977), 44, 202-6, 238; A. Michal McMahon, *The Making of a Profession: A Century of Electrical Engineering in America* (New York: The Institute of Electrial and Electronics Engineers Press, 1984), 76-7.

26 *Objects and Plan of an Institute of Technology; Including a Society of Arts, and a School of Industrial Science* (Boston: John Wilson and Son, 1861 [2nd ed.]); Francis A. Walker, "The Place of Schools of Technology in American Education," *Educational Review* (1891): 209-19; Christophe Lécuyer, "MIT, Progressive Reform, and 'Industrial Service', 1890-1920," *HSPS* 26, 1 (1995): 35-88; "The Making of a Science Based Technological University, Karl Compton, James Killian, and the Reform at MIT, 1930-1957," *HSPS* 23, 1 (1992): 153-80; Bruce Sinclair, "Inventing a Genteel Tradition: MIT Crosses the River," in *New Perspectives on Technology and American Culture*, American Philosophical Society Library Publication 12, ed. Bruce Sinclair (Philadelphia, American Philosophical Society, 1986): 1-18; Bruce Sinclair, *Philadelphia's Philosopher Mechanics: A History of the Franklin Institute, 1824-1865* (Baltimore: The Johns Hopkins University Press, 1974); Ronald Kline "Construing 'Technology' as 'Applied Science'. Public Rhetoric of Scientists and Engineers in the United States, 1880-1945," *ISIS* 86 (1995): 194-221, pp. 197-8.

27 Bruce Seeley, "Research, Engineering, and Science in American Engineering Colleges, 1900-1960," *Technology and Culture* 34 (1993): 344-86; Lécuyer, "MIT, Progressive Reform, and 'Industrial Service'."

28 See on Home Economics esp. Sarah Stage and Virginia B. Vincenti, eds., *Rethinking Home Economics. Women and the History of a Profession* (Ithaca: Cornell University Press, 1997); Margaret W. Rossiter, *Women Scientists in America: Struggles and strategies to 1940* (Baltimore: The Johns Hopkins University Press, 1982). I am grateful to Ruth Schwartz Cowan for this insight, personal communication.

29 Ronald Kline, "Construing 'Technology' as 'Applied Science'," 198-9; Entry on Thurston in *DAB*.

30 H. D. Ainsworth, *Recollections of a Civil Engineer: Experiences in New York, Iowa, Nebraska, Dakota, Illinois, Missouri, Minnesota and Colorado* (Newton, Iowa: n.p., 1901 [1893]), 187. On tensions between "school culture" and "shop-floor" culture, see Calvert's seminal book, *The Mechanical Engineer*, Chapter 8. Anecdotal evidence is provided by Spence, *Mining Engineers*, 52, 70-7. On different educational goals in electrical engineering, see McMahon, *The Making of a Profession*, 76-9.

31 G. P. C. E, "What is Wrong with the Engineering Profession?" Letter to the Editor, *Engineering News* 73, 10 (March 11, 1915): 500; see also Spence, *Mining Engineers*, 45.

32 "The College Education as a Preparation for Subsequent Specialization," *The California Journal of Technology* 3, 2 (April 1904).

33 "The Importance of Knowing How to Act," *Engineering and Contracting* 44, 17 (October 17, 1915): 321; Thomas W. Herringshaw, *Herringshaw's National Liberary of American Biography* (Chicago: American Publishers Association, 1909-1914); John W.

Leonard, *Who's Who in New York City and State Containing Authentic Biographies of New Yorkers* (New York: L.R. Hamersly, 1907); and *Who's Who in Engineering: A Biographial Dictionary of Contemporaries 1922-1923* (New York: John W. Leonard, 1922), s.v., "Gillette."

34 Ozni P. Hood, "An Apprentice System in College Shops," *Proceedings of the Society for the Promotion of Engineering Education* 5 (1899-1900).

35 Frontispiece, *California Journal of Technology* 1, 1 (February 1903). See for an excellent close reading of the sculpture, Melissa Dabakis, "Douglas Tilden's *Mechanics Fountain*: Labor and the 'Crisis of Masculinity' in the 1890s," *American Quarterly* 47, 2 (June 1995): 204-35.

36 J.A.L. Waddell, "A Proposed Study of the Spanish Language at Rensselaer," (1915) *Memoirs and Addresses of Two Decades*, ed. Frank W. Skinner (Easton, PA: Mack Printing, 1928), 310-313. Carroll W. Pursell, *The Machine in America: A Social History of Technology* (Baltimore: The Johns Hopkins University Press, 1995), 194-9.

37 National Association of State Universities and Land-Grant Colleges, *Leadership and Learning: An Interpretative History of Historically Black Land-Grant Colleges and Universities* (1993).

38 Donald Spivey *Schooling for the New Slavery: Black Industrial Education 1868-1915* (Westport, CT: Greenwood Press, 1978), Chapters 2-3; Robert McMath Jr. et al., *Engineering the New South. Georgia Tech, 1885-1985* (Athens: The University of Georgia Press, 1985), 3-35; Donald G. Nieman, *African Americans and Education in the South, 1865-1900* (New York: Garland, 1994), Introduction; David E. Wharton, *A Struggle Worthy of Note. The Engineering and Technological Education of Black Americans* (Westport, CT: Greenwood Press, 1992), 44.

39 The American Institute of Mining Engineers (AIME) organized in 1871, the American Society of Mechanical Engineers (ASME) in 1880, the American Institute of Electrical Engineers (AIEE) in 1884, the Society of Naval Architects and Marine Engineers in 1893, the American Society of Heating and Ventilating Engineers in 1894, and the American Railway Engineering Association in 1902. Layton Jr., *The Revolt of the Engineers*, Chapter 2; Bruce Sinclair, *A Centennial of the American Society of Mechanical Engineers, 1880-1980* (Toronto: University of Toronto Press, 1980), 32-4, 60-6, 113-31; McMahon, *The Making of a Profession*, 34-6, 45-6, 118-20.

40 F. H. Newell, "The Engineer Awakes," address before the American Association of Engineers, September 14, 1915, *Engineering and Contracting* (September 22, 1915): 221-3. Calculations based on data from *Who's Who in Engineering* (New York: Lewis Historical Publishing Company, 1941 [5th ed.]), xi-xxxv; Meiksins, "'The Revolt of the Engineers' Reconsidered," and "Professionalism and Conflict." Also see: Bruce Sinclair, "Local History and National Culture: Notions on Engineering Professionalism in America," *Technology and Culture* 27, 4 (October 1986): 683-93.

41 Herman K. Higgins, "Concerning the Definition of an Engineer," *Engineering News* 57, 8 (February 21, 1907); John A. Bensel, "Address at the 42d Annual Convention, Chicago, Illinois, June 21st, 1910," *Transactions of the American Society of Civil Engineers*, 70 (December 1910): 464-9; "Current Topics," *Cassier's Magazine* 37 (February 1910): 382; "What is an Engineer?" *Literary Digest* 55 (September 8, 1917): 25; editorials, letters, and articles in *Professional Engineer*, "Drafting is Engineering" (February 1920), 12; "Not

Engineers – Professional Engineers" (December 1920), 14; "Engineman" (March 1922), 17; F. L. Goodwin, "The Word 'Engineer'" and "What is an Engineer," (November 1923). See also for similar discussions by mechanical engineers in the 1890s, Calvert, *The Mechanical Engineer*, 161-3.

42 As cited in McMahon, *The Making of a Profession*, 36-7.

43 Reynolds, "Defining Professional Boundaries."

44 Meiksins, "Professionalism and Conflict."

45 Editorial, "Organization for Engineers" *The Monad* 3, 2 (February 1918): 7-10, p. 8.

46 O.H. Garman, "The Engineer's Place in Society," *The Monad* 2, 12 (December 1917): 8-9; *Who's Who in Engineering 1922-1923*, s.v., "Garman."

47 Although a minority, there is considerable literature on consultants and urban engineers: George Whipple, "What is a 'Consulting Engineer'?" *The Monad* 2, 10 (October 1917): 7-9; in mining, Spence, *Mining Engineers*, Chapter 4, esp. pp. 79, 200-3; Islam Randolph, *Gleanings from a Harvest of Memories* (Columbia: privately printed, by E. W. Stephens , 1937), 78; in mechanical engineering, Calvert, *The Mechanical Engineer*, 139-40; in civil engineering, Daniel Hovey Calhoun, *The American civil engineer: origin and conflict* (Cambridge: MIT Press, 1960), Chapter 3; in electrical engineering, McMahon, *The Making of a Profession*, 112-6. See also, Layton Jr., *The Revolt of the Engineers*, 30; on urban engineers, see: Jeffrey K. Stine, "Nelson P. Lewis and the City Efficient: The Municipal Engineer in City Planning during the Progressive Era," *Essays in Public Works History* 11 (April 1981).

48 Norman W. MacLeod, Letter to the Editor, *The Monad* 2, 2 (February 1917): 67; *Who's Who in Engineering (1922-1923)*, s.v., "MacLeod."

49 Letter to the Editor, *The Monad* 1 (May 1916): 20.

50 On engineering professionalism histories: Layton Jr., *The Revolt of the Engineers*; McMahon, *The Making of a Profession*; Calhoun, *The American Civil Engineer*; Reynolds, "Defining Professional Boundaries." For more general histories on professionalization: Robert H. Wiebe, *The Search for Order 1877-1920* (New York: Hill and Wang, 1967), 111-63 (who stresses expertise as the governing mechanism). See for interpreting this as a form of power: Mary O. Furner, *Advocacy and Objectivity* (Lexington: University Press of Kentucky, 1975); Magali Sarfatti Larson, *The Rise of Professionalism: A Sociological Analysis* (Berkeley: University of California Press, 1977); Burton Bledstein, *The Culture of Professionalism* (New York: Norton, 1976); Dorothy Ross, "Development of the Social Sciences," in *The Organization of Knowledge in Modern America, 1860-1920*, eds. Alexandra Oleson and John Vos (Baltimore: The Johns Hopkins University Press, 1979).

51 Angel Kwolek-Folland, *Engendering Business: Men and Women in the Corporate Office, 1870-1930* (Baltimore: The Johns Hopkins University Press, 1995).

52 Real Engineer, "Quacks," Letter to the Editor, *Professional Engineer* (September 1920); "The Title of Engineer," *American Machinist* 16 (1895); Calvert, *The Mechanical Engineer*, Chapter 10; Calhoun, *The American Civil Engineer*, 193; Layton *The Revolt of the Engineers*, chapters 2-3; McMahon, *The Making of a Profession*, 36-40; Spence, *Mining Engineers*, 338-9; Reynolds, "Defining Professional Boundaries."

53 For an excellent critical assessment, Meiksins, "'The Revolt of the Engineers' Reconsidered," and "Professionalism and Conflict"; Sinclair, "Notions on Engineering Pro-

fessionalism in America"; Andrew Abbott, *The Systems of Profession: An Essay on the Division of Expert Labor* (Chicago: Chicago University Press, 1988) is an important contribution to the sociological literature. See also Edwin Layton Jr., *The Revolt of the Engineers* and his essay, "Science, Business and the American Engineer," in *Engineers and the Social System*, 51-72, p. 62, fn. 4.

54 Robert W. Shelmire, "The Drafting Room," *The Monad* 2, 11 (November 1917): 13-4. Also: "General Interest," *Draftsman* 1, 6 (1902): 215. Ellen DuBois, ed., "Spanning Two Centuries: The Autobiography of Nora Stanton Barney," *History Workshop* 22 (1986): 131-52, p. 150.

55 W.B. Gump, "Why Have Technical Men Not Been Recognized As They Should Be?" *The Monad* 2, 10 (October 1917), 11-3.

56 Gump, "Why Have Technical Men," 11-13.

57 Calvert, *The Mechanical Engineer*, 149-51.

58 For glimpses of earlier forms of discontent, see: Calhoun, *The American Civil Engineer*, Chapter 6; Calvert, *The Mechanical Engineer*; Sinclair, *A Centennial of the American Society of Mechanical Engineers*, 67-9; and Spence, *Mining Engineers*, 179, 345.

59 Robert W. Shelmire, *The Draftsman* (Chicago: Chicago Publishing Bureau, 1919), 9.

60 Letter to the Editor, "The Engineer's Compensation and His Attitude Toward the Profession," *Engineering News* 61 (May 20, 1909): 553-4.

61 "Is the Draftsman a Worker at a Trade or a Professional Man?" *The Draftsman's Bulletin* 1, 6 (1901): 6. "The Modern Mechanical Drawing Room"; Editorial *The Draftsman's Bulletin* 1, 19 (1901): 1.

62 Many states did not pass laws until the late 1930s and 1940s. In 1937, the first year statistics were gathered, about 47,000 engineers were registered, and about 92,000 by 1945. *United States Directory of Registered Professional Engineers: A Biennial Publication* (New York: Lewis Historical Publishing Company).

63 Editorial, *Engineering News* 61 (May 20, 1909): 554.

64 G. P. C. E., "What is Wrong with the Engineering Profession?", 500.

65 Blank and Stigler, *The Demand and Supply*, 124, 134-6, and Tables A-10 and A-11. For figures over time and by specialization, see p. 132. From 1929 to 1946, mining engineers commanded the highest base salary, while chemical engineers received a high base pay in 1929 but the lowest pay at the end of the period. In civil engineering, the base salaries were the most stable in the inter-war period thanks in part to the federal public works projects. For comparisons of civil engineers' income to that of other occupations, Raymond H. Merritt, *Engineering in American Society, 1850-1875* (Lexington: The University Press of Kentucky, 1969), 113-7; see also, Spence, *Mining Engineers*, 167-170, 304-16.

66 Blank and Stigler, *The Demand and Supply*, 24-5.

67 M.Y. Crowdus, "The Odious," *Professional Engineer* (December 1921): 15. Editorial, "Salaries of Engineers," *The Monad* 2, 10 (October 1917): 16; "Special Experts: Charwomen and Chemists," *Chemical and Metallurgical Engineering* 4, 11 (1921): 415; also Walter Ferguson, "Compensation of Chemists and Salesmen," *Chemical and Metallurgical Engineering* 24, 11 (1921): 460.

68 Letter to the Editor, "Professional Obligations of the Engineer," *Engineering News-Record* 83, 18 (October 30-November 6, 1919): 828; W. W. K. Sparrow, "Unionism in the Profession," *Professional Engineer* (April 1920): 14-5.

69 As cited in Pursell, *The Machine*, 255-6

70 "In the Ditch!" *New Outlook* 163 (February 1934): 35-6.

71 Nora Stanton Barney, Letters to the Editor, "Industrial Equality for Women," *N.Y. Herald-Tribune* (April 21, 1933) and "Wages and Sex," *N.Y. Herald-Tribune* (July 21, 1933).

72 Meiksins, "Professional Autonomy," and "Engineers in the United States"; Nora Blatch Barney to Thomas R. Sullivan, December 21, 1944, Nora S. Blatch Papers, Courtesy of Rhoda Barney Jenkins, Greenwich, CT.

73 Alfred P. Sloan, Jr., *Adventures of a White-Collar Man* (New York: Doubleday, Doran & Company, 1941); Ruth Oldenziel, "Boys and Their Toys: The Fisher Body Craftsman's Guild, 1930-1968 and the Making of a Male Technical Domain," *Technology and Culture* 38, 1 (January 1997): 60-96.

74 Bensel, "Address at the 42d Annual Convention."

75 John J. Carty, "Ideals of the Engineer," address delivered at the Presentation of the John Fritz Medal at A.I.E.E. Convention, New York, NY, February 15, 1928, *Journal of the American Institute of Electrical Engineers* 47, 3 (March 1928): 210-2. See, for his earlier position, when he tried to exclude managers from the field of electrical engineering, Spence, *Mining Engineers*, 111.

76 Great Engineers and Pioneers in Technology I, eds. Ronald Turner and Steven L. Goulden (New York: St. Martin's Press, 1981), 413-4. For an early assessment of the appropriateness of the definition, see Editorial, *Engineering News* 52, 1 (July 7, 1904): 8; McMahon, *The Making of a Profession*, 108-9.

CHAPTER 3

1 This chapter is based on a list compiled by Louis Kaplan, *A Bibliography of American Autobiographies* (Madison: University of Wisconsin Press, 1963) to which other entries have been added, culled from library collections, secondary literature, and a shelf-list search at the Library of Congress. The number of autobiographies discussed in this chapter is by no means exhaustive as new material continues to surface, but it provides a rough estimate of the literary production of engineers. On producing culture while at work, see: Cynthia Cockburn, *Machinery of Dominance: Women, Men, and Technical Know-how* (Boston: Northeastern University Press, 1988), 167-97.

2 Secondary literature on the writing of autobiographical narratives is vast, varied, and very uneven in quality. It has preoccupied historians, psychoanalysts, and literary critics alike. Much confusion and controversy has centered around the similarities and differences between biographies and autobiographies. My understanding of autobiographies is derived from a multitude of secondary writings on the subject, of which the following I have found the most useful: Jan Romein, *De Biografie: een inleiding* (Amsterdam: Ploegsma, 1946); Paul de Man, "Autobiography as De-facement," *MLN* 94, 5 (December 1979): 919-30; Daniel Berteaux, *Biography and Society: The Life History Approach in the Social Sciences* (Beverly Hills: SAGE Publications, 1981); Avrom Fleishman, *Figures of Autobiography: The Language of Self-Writing in Victorian and Modern England* (Berkeley: University of California Press, 1981); Estelle C. Jelenik, ed., *Women's Autobi-*

ographies: Essays in Criticism (Bloomington: Indiana University Press, 1980); Sidonie Smith, *A Poetics of Women's Autobiography. Marginality and the Fictions of Self-Representation* (Bloomington: Indiana University Press, 1987). The journal *Biography* is devoted to specialized topics in this field.

3   John Kotre, *Outliving the Self: Generativity and the Interpretation of Lives* (Baltimore: The Johns Hopkins University Press, 1984), 10. Kotre's work elaborates on Erik Erikson's and Daniel Levinson's notion of generativity and creativity. His study is particularly salient for white middle-class men such as those discussed here. Carol Gilligan has severely and effectively challenged the male conceptualization of this body of literature. While historically valid, both approaches suffer from universalist claims. Erik H. Erikson, *Childhood and Society* (New York: W. Norton, 1950); Daniel Levinson et al., *The Seasons of Man's Life* (New York: Alfred Knopf, 1978); Carol Gilligan, *In a Different Voice: Psychological Theory and Women's Development* (Cambridge: Harvard University Press, 1982), 150-5.

4   In the tradition of the work of Samuel Smiles (1812-1904), *Self-Help* (London, 1859) and *Selections from Lives of the Engineers, with an Account of their Principal Works*, ed. Thomas Parke Hughes (Cambridge: MIT Press, 1966). On Smiles' inspiration of engineers, see for example: Onward Bates, "Onward and Onward: A Romance of Four Continents. Memorabilia of an Engineer" (mimeo by Ruth I. Hill, 1933), 15. For a general assessment of Smiles' work, see: Tim Travers, *Samuel Smiles and the Victorian Work Ethic* (New York: Garland Publishing, 1987 [1970]).

5   Embury A. Hitchcock, *My Fifty Years in Engineering* (Caldwell, Idaho: Caxton printers, 1939), Dedication. The following autobiographies directly addressed young male adolescents: John Frank Stevens, *An Engineer's Recollections* (New York: McGraw-Hill, 1936), xiii, 66; William LeRoy Emmet, *The Autobiography of an Engineer* (Albany: Fort Orange Press, 1931), foreword and xi-xii; Albert Sauveur, *Metallurgical Reminiscences* (New York: American Institute of Mining and Metallurgical Engineers, 1937), iii; "Selecting and Training of Men," in Paul Weeks Litchfield, *Autumn Leaves: Reflections of an Industrial Lieutenant* (Cleveland: The Corday & Gross, 1945).

6   Leigh Bristol-Kagan, "Chinese Migration to California, 1851-1882: Selected Industries of Work, the Chinese Institutions and the Legislative Exclusion of a Temporary Labor Force" (Ph.D. diss., Harvard University, 1982), 21-31.

7   Michael I. Pupin, *From Immigrant to Inventor* (New York: Scribners, 1923); Benjamin G. Lamme, *Electrical Engineer. An Autobiography* (New York: Putnam, 1926); Emmet, *Autobiography*; Litchfield, *Autumn Leaves*.

8   James M. Searles, *Life and Times of a Civil Engineer; Supplemented by the True Theory of the Mississippi River* (Cincinnati: Robert Clarke, 1893), 3, 21. In his preface and appendix, Searles used excerpts from Charles Ellett Jr.'s 1857 report to Congress on the Mississippi River to support his argument for an integrated water system which would prevent severe flooding. The argument for a comprehensive plan rather than further strengthening the levees represented a minority position in 1893. Robert W. Harrison, *Levee Districts and Levee Building in Mississippi: A Study of State and Local Efforts to Control Mississippi River Floods* (1951), 121-41. On the role of the levee system in the Southern economy, see Robert L. Branfon, *Cotton Kingdom of the New South: A History of the*

Yazoo – Mississippi Delta from Reconstruction to the Twentieth Century (Cambridge, MA: Harvard University Press, 1967), 1-38.

9   Searles tried to convince both Deer Creek planters and Vicksburg merchants of the feasibility and profitability of the Memphis and Vicksburg Railroad Company, Searles, *Life and Times*, 91-2. Except for his autobiography, no information appears in specialized engineering biographical dictionaries, biographical sources of the Mississippi and Louisiana region where Searles worked for the greater part of his life, or in Congressional records. I am grateful to Nancy Bercaw, however, for bringing to my attention two references in the *Greenville Times* (Mississippi) which places him in Greenville in 1874 as part of a United States Army Levee investigation. ASCE Application Register, *Greenville Times* (5 August and 12 September 1874).

10 Searles, *Life and Times*, 4.

11 Searles, *Life and Times*, 110-12. Irish immigrants were recruited for levee work, Robert W. Harrison, "Early State Flood-Control Legislation in the Mississippi Alluvial Valley," *Journal of Mississippi History* 23, 2 (April, 1961): 104-26, p. 125.

12 Peter F. Meiksins, "Science in the Labor Process: Engineers as Workers," in *Professionals as Workers: Mental Labor in Advanced Capitalism*, ed. Charles Derber (Boston: G. K. Hall, 1982), 121-40; Alfred D. Chandler Jr., *The Visible Hand: The Managerial Revolution in American Business* (Cambridge, MA: Harvard University Press, 1977), 2.

13 Searles, *Life and Times*, 97. James L. Roark, *Masters without Slaves: Southern Planters in the Civil War and Reconstruction* (New York: Norton, 1977), Chapter 5. His involvement in engineering before the Civil War suggests his family was already in economic decline before Emancipation. I am grateful to Joe Reidy for sharing his thoughts with me on this issue. Emmet similarly could put a price on his labor and the profit he made for his employers, expressing his frustration at being undervalued, Emmet, *Autobiography*, 94.

14 Clark C. Spence, *Mining Engineers & the American West: The Lace-Boot Brigade, 1849-1919* (New Haven: Yale University Press, 1970), 98-103; John Fritz, *The Autobiography of John Fritz* (New York: John Wiley & Sons, 1912), 99-107.

15 See Branfon, *Cotton Kingdom*, 1-38; Robert W. Harrison, *Levee Districts and Levee Building in Mississippi*, 130 and "Early State Flood-Control Legislation", 104, 117.

16 After the Civil War, many Southern men went into engineering in order to replenish their finances and tried to reconstitute their (problematic) Southern identity as a professional one, with varying degrees of success. Wilson and Searles were the most successful at this transformation. Pickett, however, wrote two autobiographical narratives: an engineering autobiography signalling his allegiance to white northern manhood, and the other, a hunting autobiography, suggesting his loyalty to Southern white manhood. Robert Milton Howard, *Reminiscences* (Columbus, GA.: Gilbert press, 1912); William D. Pickett, *A Sketch of the Professional Life of William D. Pickett of Four Bear, Wyoming* (Louisville: John P. Morton, 1904) and his "Memories of a Bear Hunter," in *Hunting at High Altitudes: The Book of the Boone and Crockett Club*, ed. George Bird Grinnell (New York: Harper and Brothers, 1913). Frank E. Smith, *The Yazoo River* (Jackson: University Press of Mississippi, 1954), 217-42; William H. Wilson, *Reminiscences*, ed. Elizabeth B. Pharo (Philadelphia: Patterson and White, 1937). See also: Isham Randolph, *Gleanings from a Harvest of Memories* (Columbia: E.W. Stephens,

1937); Richard Justin McCarty, *Work and Play* (Kansas City: Empire Press, 1912); Matthew Walton Venable, *Eighty Years After: or Grandpa's Story* (Charleston: Hood-Heiserman-Brodhag, 1929).

17 Wilson, *Reminiscences*, 1. John D. Littlepage with Demaree Bess, *In Search of Soviet Gold* (New York: Harcourt, Brace and Company, 1938), xii-xiii; Emmet, *Autobiography*, xiv; Charles T. Porter, *Engineering Reminiscences* (New York: Wiley and Sons, 1912), preface.

18 Robert Ridgway with Isabelle Law Ridgway, *Robert Ridgway* (New York: privately printed, 1940), Preface; R. W. Lawton, *An Engineer in the Orient* (Los Angeles: Walton & Wright, 1942), Introduction. See for examples of the technical knowledge and eloquence contained in visual representations: Engineering Notebooks Collection, Archives Center, National Museum of American History, Smithsonian Institution, Washington, DC. On the communicative power of visual language, Embury A. Hitchcock, a mechanical engineer who "carried on quite a conversation [on air brakes] by means of a notebook and pencil and look and nods," with a French colleague at the 1900 Paris exposition, without either of them knowing the other's native tongue; Hitchcock, *Fifty Years*, 154. Also Eugene S. Ferguson's classic article: "The Mind's Eye: Nonverbal Thought in Technology," *Science* (August 26, 1977): 827-36; Harold Belofsky, "Engineering Drawing – a Universal Language in Two Dialects," *Technology and Culture* 32, 1 (January 1991): 23-46; Steve Lubar, "Representing Technological Knowledge," (paper presented at the Society of the History of Technology Conference, Madison, October 1991).

19 J.A.L. Waddell, "The Engineer's English," and "Technical Book Writing," both in *Memoirs and Addresses of Two Decades by J.A.L. Waddell, Consulting Engineer*, ed. Frank W. Skinner (Easton, PA.: Mack Printing, 1928): 355-9, p. 356 and 325-53. See also: Editorials in *California Journal of Technology* 1, 2 (April 1903): 100-1; 1, 3 (November 1903); Thomas A. Rickard, *A Guide to Technical Writing* (San Francisco: Mining and Scientific Press, 1908) and *Technical Writing* (New York: John Wiley & Sons, 1920); Hitchcock, *Fifty Years*, 169-70, 240.

20 Danforth H. Ainsworth, "Discussion on Railroad Location," *Transactions of the American Society of Civil Engineers* 31 (January-June 1894): 95-8, p. 97.

21 Daniel M. Barringer to Guy P. Bennett in 1899, as quoted in Spence, *Mining Engineers*, 89.

22 George W. Brown, "Private Log Book of a little trip to Nicaragua, 1897-8," Division of Engineering and Industry, National Museum of American History, Smithsonian Institution, Washington, DC, (E&I, NMAH, SI, hereafter). For other examples, see: November 3, 1880, Nathaniel Chapin Ray correspondence to his sister Anna, New Haven Historical Society, New Haven, CT., (Ray Papers, hereafter); Cecile Hulse Matschat, *Seven Grass Huts: An Engineer's Wife in Central-and-South America* (New York: Farrar & Rinehart, 1939), 133; Ainsworth, *Recollections*, 172-3; Herman Haupt, "Reminiscences of Early History of the Pennsylvania Railroad Company," typescript, Division of Engineering and Industry, Biographical File, NMAH, SI, 3; Littlepage, *In Search of Soviet Gold*, xii-xiii: Hitchcock, *My Fifty Years*, 60, 240.

23 Onward Bates used this third-person narrative in his autobiography as a rhetorical device, *Onward and Onward*, 3; Wilson, *Reminiscences*, 1; Littlepage, *In Search of Soviet Gold*, xii-xiii; Emmet, *Autobiography*, xiv.

24 Ainsworth, *Recollections*, 12, 31, 187. Ainsworth, a member of the ASCE from 1886 until his death in 1904, appears only once the organization's annals with an obituary: *Proceedings of the American Society of Civil Engineers* 31 (1905), 241-2.

25 For examples of their occupational writing, see: John L. Pott, Orchard Iron Works day book, July 6, 1861-1862, MSS Collection, #1990.0178; G.S. Morison Engineers' Logbooks, 1863-1903, MSS Collection both in Div. of E&I, NMAH, SI; Engineers' Notebook Collection, Archives Center, NMAH, SI, Washington, DC. The diary of Erasmus D. Leavitt, intended as an engineers' log, was exceptional in evolving into a personal record. Erasmus D. Leavitt Jr. Collection, E&I, NMAH, SI.

26 Henry Root, *Henry Root, Surveyor, Engineer, and Inventor* (San Francisco: privately printed, 1921), 7.

27 Littlepage, *In Search of Soviet Gold*, xv.

28 Matschat, *Seven Grass Huts*, 54. Obituary, *The New York Times* (March 10, 1976): 42. I am grateful to Jeffrey K. Stine for calling my attention to Matschat's autobiography. On gendered differences of the frontier landscape, see: Annette Kolodny, *The Land before Her: Fantasy and Experience of the American Frontiers, 1630-1860* (Chapel Hill: The University of North Carolina Press, 1982), 35-54.

29 Ainsworth, *Recollections*, 33. See for a detailed discussion, JoAnne Yates, *Control through Communication: The Rise of System in American Management* (Baltimore: The Johns Hopkins University, 1989), 252, 232; Spence, *Mining Engineers*, 99. On the use of efficient language, see: Cecilia Tichi, *Shifting gears: Technology, Literature, Culture in Modernist America* (Chapel Hill: The University of North Carolina Press, 1987).

30 Ainsworth, *Recollections*; Littlepage, *In Search of Soviet Gold*; Charles T. Porter, *Engineering Reminiscences* (New York: Wiley and Sons, 1912); Ridgway, *Robert Ridgway*; Searles, *Life and Times*; and Wilson, *Reminiscences*.

31 Fritz, *Autobiography*, 153, and 156, 168, 171.

32 Emmett, *Autobiography*, 86-92, 117.

33 Searles, *Life and Times*, 73. Donald Yacovone, "Abolitionists and the Language of Fraternal Love," in *Meanings for Manhood*, eds. Mark C. Carnes and Clyde Griffen (Chicago: University of Chicago Press, 1990), 85-95.

34 Ainsworth, *Recollections*; Emmet, *Autobiography*; Porter, *Engineering Reminiscences*; Ridgway, *Robert Ridgway*; Root, *Surveyor, Engineer*.

35 Bates, *Onward and Onward*, 3; see also *Bates et al. of Virginia and Missouri* (Chicago: printed for private circulation, 1914), 155.

36 Alfred West Gilbert, *Colonel A. W. Gilbert; Civil Engineer* (Cincinnati: Historical and Philosophical Society of Ohio, 1934), 25; Obituary, *The New York Times* (August 18, 1938): 19; 1. See also: Walter Holcomb, *Memories of Walter Holcomb of Torrington, Litchfield County, Connecticut with a few departures in genealogy, public records, customs etc.* (N.p., 1935), 25.

37 Fritz, *Autobiography*, 160-1, and 146, 301; Benjamin Franklin Fackenthal, "John Fritz, the Iron Master," *The Pennsylvania German Society* 34 (1929): 95-112, pp. 102, 104.

38 Bates, "Onward and Onward," 4, 20.

39 See the roles of women in the production of the autobiographies of Littlepage (initiated and produced by journalist Demaree Bess), Ridgway (instigated by, told to, and written by his second wife Isabelle Ridgway), Wilson (edited and published by his granddaughter), Bates (mimeographed by Ruth Hill); Tom Cooney, *Meet Tom Cooney* (collaborated with Mildred H. Comfort) (Minneapolis: Lund Press, 1945). See also the Nathaniel Chapin Ray papers which were preserved and donated by his sister, the author Anna Chapin Ray. His sister failed to leave any of her own papers, as did her brother. Also, see: Brown diary, "Private Log Book." Exceptional in this respect were Emmet, *The Autobiography* and Hitchcock, *Fifty Years*, 138, 140.

40 Matschat, *Seven Grass Huts*, 8, 6-7. Taciturnity returns throughout the narrative as a powerful theme, see pp. 122, 135, 158, 231.

41 James Worrall, *The Memoirs of James, Civil Engineer: with Obituary Postscript by a Friend* (Harrisburg: E. K. Meyer, 1887), 5. See also his *Report of a Survey of South Pennsylvania Railroad* (n.p.: 1862) and *Report of the Pennsylvania Board of Pennsylvania and New York Joint Boundary Commission* (Harrisburg: L.S. Hart, State Printer, 1878). On contacts with working-class manliness, see also letters September 8, 1878 and October 13, 1880, Ray papers; Spence, *Mining Engineers*, 338; Peter Way, *Common Labor: Workers and the Digging of North American Canals, 1780-1860* (Baltimore: The Johns Hopkins University Press, 1993), 14-15.

42 See for a vivid description: Thomas J. Misa, *A Nation of Steel: The Making of Modern America, 1865-1925* (Baltimore: The Johns Hopkins University Press, 1995), 1-4.

43 Isham Randolph, *Gleanings from a Harvest of Memories* (Columbia, E.W. Stephens, 1937); *DAB*, s.v., "Randolph"; Donald L. Miller, *City of the Century. The Epic of Chicago and the Making of America* (New York: Touchstone, 1996), 429-31.

44 David Montgomery, "Workers' Control of Machine Production in the Nineteenth Century," in *Workers' Control in America* (New York: Cambridge University Press, 1979), 1-31; and *The Fall of the House of Labor* (Cambridge: Cambridge University Press, 1987), Chapter 1; Herbert G. Gutman, "Two Lockouts in Pennsylvania, 1873-1874," *Pennsylvania Magazine of History and Biography* 83 (July 1959): 307-326; Gary M. Fink, ed., *Labor Unions* (Westport, CT: Greenwood Press, 1977), 158-60; L.T.C. Rolt, *A Short History of Machine Tools* (Cambridge: MIT Press, 1965), Introduction; Melissa Dabakis, "Douglas Tilden's *Mechanics Fountain*: Labor and the 'Crisis of Masculinity' in the 1890s," *American Quarterly* 47, 2 (June 1995): 204-35.

45 For example, the ASME published: *John Edson Sweet* (1925), *A Biography of Walter Craig Kerr* (1927) both by Albert W. Smith; William F. Durand, *Robert Henry Thurston* (1929); L. P. Alford, *Henry Lawrence Gantt: Leader in Industry* (1934), Joseph W. Roe, *James Harness* (1937), Mortimer E. Cooley, *Scientific Blacksmith* (1947). *National Cyclopaedia*, s.v., "Wiley"; Misa, *Nation of Steel*, 74 n. 76; Amy Sue Bix, "Inventing Ourselves Out of Jobs: America's Depression Era Debate Over Technological Unemployment," (Ph.d. Diss., The Johns Hopkins University, 1994), 459.

46 Fritz, Autobiography, 236, 241, 245, 249.

47 See also: Warren I. Susman's classic, "Culture Heroes: Ford, Barton, Ruth," in *Culture as History. The Transformation of American Society in the Twentieth Century* (New York: Pantheon Books, 1984), 122-49.

48 Arthur G. Burgoyne, *Homestead* (Pittsburgh: privately printed, 1893), 130-197; Montgomery, *House of Labor*, 36-43; David Brody, *Steelworkers in America* (New York: Harper Torchbooks, [1960]); Jeremy Brecher, *Strike!* (Boston: South End Press, 1972), 53-63; Fink, ed., *Labor Unions*, 160.

49 On the functions of jokes, see: Mary Douglas and Robert Darnton in *Rethinking Popular Culture: Contemporary Thought in Cultural Studies*, eds. Chandra Mukerji and Michael Schudson (Berkeley: University of California Press, 1991), 291-310.

50 Fritz, *Autobiography*, 296-7. *DAB*, s.v., "Rossiter." Philip S. Foner, *The Policies and Practices of the American Federation of Labor, 1900-1909* (New York: International Publishers, 1964), esp. chapter 4; John W. Hevener in *Labor Unions*, 412-13; Bruce Sinclair, *A Centennial of the American Society of Mechanical Engineers, 1880-1980* (Toronto: University of Toronto Press, 1980), 65-7, 85, 98, 128-9, 184-6.

51 Fritz, *Autobiography*, 278-9; A female perspective on the celebration was provided anonymously by "The Banquet as Seen by One in the Gallery," who wryly commented that "all human history attests; That happiness for man, the hungry sinner; Since Eve ate the apples, much depends on dinner!", pp. 313-18; S.v., *DAB* "Fritz."

52 Fritz, *Autobiography*, 22, 33, 62, 82, 100, 129, 126-27, 132.

53 Fritz, *Autobiography*, 275, 311.

54 Fritz, *Autobiography*, 53-4, 205, 28-9, 33-34. See for other examples, Hitchcock, *Fifty Years*, 48, 82, 98, 107, 111 (jokes); Ray Papers, letters October 31, 1878 and November 31, 1878 (on jokes); September 9, 1878 (language).

55 Fritz, *Autobiography*, Chapter 20, "Puddling."

56 Robert W. Shelmire, *The Draftsman* (Chicago: Chicago Publishing Bureau, 1919), 10-11, 33; Peter F. Meiksins, "Professionalism and Conflict: The Case of the American Association of Engineers," *Journal of Social History* 19, 3 (Spring 1986): 403-421, p. 406; Fink, *Labor Unions*.

57 Edward Carpenter, "Tomorrow," *The Draftsman* (Cleveland) 2, 4 (April 1903).

58 A.J.S., "The Draftsman," *The Draftsman* (Philadelphia) 1, 6 (January 1928): 16.

59 See also Misa, *Nation of Steel*, 174.

60 Samuel W. Traylor, *Out of the SouthWest: A Texas Boy* (Allentown, PA: Schlicher & Son, 1936), Foreword; John G. Cawelti, *Apostles of the Self-Made Man* (Chicago: University of Chicago Press, 1965).

61 Gilbert, *Colonel A. W. Gilbert*, Foreword.

62 Paul Starret, *Changing the Skyline* (New York: Wittlesley House, 1938), 316.

63 Stevens, *Recollections of an Engineer*, 65; see also J. F. Stevens to Professor Carpsis of Manila, March 27, 1936, Dibner Library, NMAH, SI, Washington, DC.

64 David McGee, "Making Up Mind: The Early Sociology of Invention," *Technology and Culture* 36, 4 (October 1995): 773-801; Bix, "Inventing Ourselves Out of Jobs," 458-9; *DLB* 46, s.v., "McGraw-Hill"; Arthur P. Molella, "The First Generation: Usher, Mumford, and Giedion," in *In Context: The History and the History of Technology*, eds. Stephen H. Cutliffe and Robert C. Post (Bethlehem: Lehigh University Press, 1989): 88-105. On collective memory: Maurice Halbwachs, *The Collective Memory*, Intro. Mary Douglas, trans. Francis J. Ditter and Vida Yazdi (New York: Harper & Row, 1980 [1950]); Philip Nora, "Between Memory and History," *Memory and Counter-Memory*. Special Issue, eds. Nathalie Zemon Davis and Rudolph Starn, *Representa-*

tions 26 (1989): 7-22; David Thelen, "Memory and American History," *Journal of American History*, Special Issue 75, 4 (March 1989): 1117-29.
65 Gilbert, *Colonel A. W. Gilbert*, Introduction.

CHAPTER 4

1   Cyrus Townsend Brady, *Web of Steel* (New York: Fleming H. Revell, 1916), Preface.
2   *The New York Sun* (January 16, 1916): 7; Julian Chase Smallwood, "Engineering and Art," *Cassier's* 37 (January 1910): 213-6; Louis H. Gibson, "Art and Engineering," *Scientific American Supplement* 58 (October 1904): 24033-4; James P. Haney, "Art and the Engineer: Combining the Useful and the Beautiful," *Scientific American Supplement* 71 (February 25, 1911): 119; "Artistic Bridge and Highway Railings," *American City* 20 (March 1919): 234-39; "Structural Art," *Literary Digest* 67 (December 25, 1920): 26-7. See also for a similar trend in the Netherlands, Harry Lintsen's *Ingenieur van beroep: historie, praktijk, macht en opvattingen van ingenieurs in Nederland* (Den Haag: Ingenieurspers, 1985), 12-3.
3   E.A. Van Deusen, "The Engineer's Place in Fiction," *Professional Engineer* (September 1922): 15.
4   J.H. Prior, "Potential Good of A.A.E. Beyond Recognition," *The Monad* (December 1916): 11.
5   The dissemination of the figure of the engineer occurred both through popular fiction, the progress reports on engineering projects appearing in the daily press (e.g. *The New York Sun*) and genteel periodicals including *McClure's, Scribners', Harper's*, and *The Century*, all of which frequently featured articles on engineers starting in the 1890s.
6   Echoing Hawthorne, Ann Douglas presented the classic argument in *The Feminization of American Culture* (New York: Alfred A. Knopf, 1977), which has been critically revised: Sandra M. Gilbert and Susan Gubar, *No Man's Land: The Place of the Woman Writer in the Twentieth Century. The War of the Words* (New Haven: Yale University Press, 1988); Janice A. Radway, *Reading the Romance: Women, Patriarchy, and Popular Literature* (Chapel Hill: The University of North Carolina Press, 1984); Cathy N. Davidson, *Revolution and the Word: The Rise of the Novel in America* (New York: Oxford University Press, 1986), 45-9, 112-25; Barbara Sicherman, "Sense and Sensibility: A Case Study of Women's Reading in Late-Victorian America," in *Reading in America: Literature and Social History*, ed. Cathy N. Davidson (Baltimore: The Johns Hopkins University Press, 1989), 201-25; see also T.J. Jackson Lears, *No Place of Grace: Antimodernism and the Transformation of American Culture, 1880-1920* (New York: Pantheon, 1981), 103-4.
7   Unless otherwise indicated in the text, this chapter is based on the following: Harold Bell Wright, *The Winning of Barbara Worth* (Chicago: Book Supply, 1912); John Fox Jr., *The Trail of the Lonesome Pine* (New York: Scribner's Sons, 1908); Zane Grey, *The U.P. Trail* (New York: Harper and Brothers, 1918). Francis Lynde, *The Quickening* (Indianapolis: Bobbs-Merrill, 1906); Frank H. Spearman, *Whispering Smith* (New York: Charles Scribner's Sons, 1906); James Oliver Curwood, *The Danger Trail* (New York: Grosset and Dunlap, 1910); Rex Beach, *The Iron Trail: An Alaskan Romance* (New

York: Harper and Brothers, 1913); H. Irving Hancock, *The Young Engineers in Nevada* (Philadelphia: Henry Altemus, 1913); The work of Honoré (McCue) Willsie Morrow, *The Enchanted Canyon* (New York: A. L. Burt, 1921), *Still Jim* (New York: Frederick A. Stokes, 1914) and *The Heart of the Desert* (New York: A.L. Burt, 1913); Brady, *Web of Steel*; Francis Arnold Collins, *The Fighting Engineers: The Minute Men of Our Industrial Army* (New York: Century, 1918); Edward Halsey Foster, "This Cowboy Is an Engineer: Popular Fiction, Politics and the West," *Rendezvous* 19, 1 (Fall 1983): 1-7.

8 Cecelia Tichi noted, but did not elaborate on, the manliness in the representation of engineers in her well-researched and beautifully written book, *Shifting Gears: Technology, Literature, Culture in Modernist America* (Chapel Hill: University of North Carolina Press, 1987), 99. I especially benefitted from Elizabeth Ammons's article, "The Engineer as Cultural Hero and Willa Cather's First Novel, *Alexander's Bridge*," *American Quarterly* 38 (Winter 1986): 746-60, in which she deftly treats the issue of masculinity in Cather's novel. Theodore Ziolkowski, "The Existential Anxieties of Engineering," *The American Scholar* 53 (Spring 1984): 197-218.

9 See for the *locus classicus* the doctrine of the strenuous life, "of toil and effort, of labor and strife; to prevail that the highest form of success which, not from the man who desires mere easy peace, but the man who does not shrink from danger, from hardship, or from bitter toil, and who out of these wins the splendid ultimate triumph." Theodore Roosevelt, "The Strenuous Life: Speech Before the Hamilton Club," April 10, 1899, in *The Strenuous Life* (New York: Century, 1902 [1899]), 1-21. The argument is presented by Foster in his article "This Cowboy Is an Engineer", but is implicitly refuted by Tichi who sees the engineer as foremost a ruling-class figure and professional in the midst of a civilizing effort. Foster concentrates on the western novels within the genre, while Tichi's analysis focuses on writers from the eastern establishment. Tichi, *Shifting Gears*, 117-20, and esp. 123. See also David B. Davis, "Marlboro Country," in *From Homicide to Slavery: Studies in American Culture* (New York: Oxford University Press, 1987 [1979]), 104-12.

10 Richard Harding Davis, *Soldiers of Fortune* (New York: Century, 1897), 13-4; James R. McGovern, "David Graham Philips and the Virility Impulse of the Progressives," *New England Quarterly* 39 (1966): 334-55; Lears, *No Place of Grace*.

11 Annette Kolodny, *The Land Before Her: Fantasy and Experience of the American Frontiers, 1630-1860* (Chapel Hill: The University of North Carolina Press, 1984); Brady, *Web of Steel*; Wright, *The Winning of Barbara Worth*. Their Dutch counterparts socialized in comfortable bourgeois settings fully inhabited by important female characters. In American representations, the American West posed a challenge; whereas in the Dutch novels, *female fatales* threatened the engineers' road to success. For examples, see: Busken Huet, *Lidewijde* (1868); P.A. Daum, *H. van Brakel, ing. B.O.W.: oorspronkelijke roman door Maurits* (Leiden: A.W. Sijthoff, 1890); Cornélie Huygens, *Barthold Meryan* (Amsterdam: P. N. van Kampen & Zoon, n.d.).

12 The threat was most clearly formulated in Wright's *When a Man's a Man* (1916), as Fritz H. Oehlschlaeger argues in "Civilization as Emasculation: The Threatening Role of Women in the Frontier Fiction of Harold Bell Wright and Zane Grey," *The Midwest Quarterly* 22, 4 (Summer 1981): 346-60. John P. Ferre, *A Social Gospel for Millions: The Religious Bestsellers of Charles Sheldon, Charles Gordon, and Harold Bell Wright* (Bowling

Green, OH: Bowling Green University Popular Press, 1988), 12, 56-68; Edward Ifkovic in *DLB* 9, s.v., "Wright."

13 Christopher P. Wilson, "The Rhetoric of Consumption: Mass-market Magazines and the Demise of the Gentle Reader, 1880-1920," in *The Culture of Consumption*, eds. Richard Wightman Fox and T.J. Jackson Lears (New York: Pantheon Books, 1983), 39-64, pp. 50-56.

14 Charles Cantalupo in *DLB* 19, s.v., "Kipling"; Mary A. O'Toole in *DLB* 34, s.v., "Kipling."

15 On Kipling's engineering stories see Jean-François Orjollet, "Individu, type, règle: Kipling et ses *engineers*," *Cahiers Victoriens et Edouardiens* 18 (Novembre 1983): 59-67; Christopher Harvie, "'The Sons of Martha': Technology, Transport and Rudyard Kipling," *Victorian Studies* 20 (1977): 269-82.

16 "Kipling Writes Poem for Technical Magazine," *The New York Times* (March 6, 1935): 17; R. Kipling, "Hymn of Breaking Strain," *Civil Engineering* (1935); "Romance of Engineering," *Scientific American* 106 (February 10, 1912): 124.

17 R[alph] W[aldo] Lawton, *An Engineer in the Orient* (Los Angeles; Walton & Wright, 1942). Obituary, *The New York Times* (August 22, 1943), 36:5; see also Walter Holcomb's use of Kipling, *Memories of Walter Holcomb of Torrington, Litchfield County, Connecticut with a Few Departures in Genealogy, Public Records, Customs etc.* (N.p., 1935), 47.

18 C.E. Moorhouse, "Mr Wardrop's Problem: Excerpts from a Talk on Kipling and Technology," *Kipling Journal* 61 (March, 1987): 10-22, p. 10.

19 Robert T. Gebler, "The Engineer (with Apologies to Kipling)," *Engineering News* 73, 2 (January 14, 1915): 17.

20 I am especially grateful to Matt L. Turner with whom I collaborated on a close reading of the poem. See: "Contesting the Good Part: Kipling's Sons of Martha and Definitions of Labor" (paper presented by Matt L. Turner and Ruth Oldenziel at the Department of Comparative Literature Seminar, Yale University, April 11, 1990). The poem first appeared in the following dailies: *The Standard* (London), April 29, 1907; *Associated Sunday Magazines* (April 28, 1907); *The New York Tribune* (April 28, 1907), illustrated by Winter; *Philadelphia Press* (April 28, 1907). In pamphlet form, it appeared in the following editions: "The Sons of Martha," (New York: Doubleday, Page, 1907.) Broadside. Description: Title within ornamental border. [Photostat copy]; "The Sons of Martha," (Garden City: Doubleday, Page & Company, 1907). Broadside. Description: First American edition. Livingston Supplement, 151; "The Sons of Martha," New York, 1907. Copy detached from *Sunday Magazine* for April 28, 1907. No detailed critical analysis exists of the poem. Despite the title "The Sons of Martha," Harvie's article deals not with the poem but with Kipling's views on technology and society.

21 Kipling's reputation with the critics was declining, but this poem was well-received: "The Sons of Martha," *The Independent* (New York) 62 (May 23, 1907): 1217-8; Lyman Abbott, "The Book Table: Devoted to Books and Their Maker. Kipling's Latest Word," *Outlook* 123 (Sept. 24, 1919): 144. F.H., "Literature. Recent Poetry," *The Nation* 109, 2821 (July 26, 1919): 115-7; "Volumes of Poetry," *The American Review of Reviews* 60 (September 1919): 446-7; Charles Hanson Towne, "The Vanished Yeats, The Never-Vanishing Kipling," *Bookman* 49 (1919): 617-2; The reviewer in the *Dial* found

Kipling politically repulsive because "a mind so obviously wrong about women (The Female of the Species) will probably [be?] distrusted when it considers labor (The Sons of Martha)..." yet believed that the poem offered a rare glimpse of the author's old power, "Notes on New Books," *The Dial* (May 31, 1919): 57.

22 H.M., "Kipling as a Poet," *Poetry* 48, 1 (April 1936): 32-6.

23 Dixon Merritt, *Sons of Martha. A Historical and Biographical Record Covering a Century of American Achievement by an Organization of Master Builders* (New York: Mason and Hanger Company, 1928).

24 "The Sons of Martha," *The New York Times*, editorial (September 23, 1938): 26:2; Albert J. Franck, Letter to the Editor, *The New York Times* (September 28, 1938): 24:7.

25 Augustine J. Frederich, ed., *Sons of Martha: Civil Engineering Readings in Modern Literature* (New York: American Society of Civil Engineers, 1989).

26 Marion Harland, "Martha and Her American Kitchen," *Woman's Home Companion* (March 1905): 12-13, 51, 55. See also: Elizabeth Foote, "Girl of the Engineers," *Atlantic Monthly* 95 (March 1905): 381-91; "From a Sister of Martha," *Atlantic* 27 (March 1921): 424-6.

27 "The Sons of Martha by Rudyard Kipling," (Chicago: printer John F. Higgins, n.d.). According to Harvie, by 1912 Kipling "had come to fear the growing confrontation between labour and management." Harvie, "The Sons of Martha," 278.

28 Philip S. Foner, *History of the Labor Movement in the United States. Post-War Struggles, 1918-1920* (New York: International Publishers, 1988).

29 G.S.B., "The Sons of Mary," *Literary Digest* 63 (November 8, 1919): 39.

30 On Kipling's attitude toward unions, see Harvie, "Sons of Martha,"; on modern artists' representation of themselves as designers, see Tichi, *Shifting Gears*, 173.

31 Surprisingly, T.S. Eliot was in large part responsible for restoring Kipling's reputation as a respectable writer. By examining the most frequently quoted and popular writer of the late nineteenth and early twentieth centuries, Eliot was able to define his own modernist agenda with greater clarity and precision. To him, Kipling represented everything he was not. T.S.E[liot], "Reviews, Kipling Redivivus," *Athenaeum* (May 9, 1919): 288-9, p. 289.

32 Mary Hallock Foote, *A Victorian Gentlewoman in the Far West: The Reminiscences of Mary Hallock Foote*, ed. Paul Rodman (San Marino: Huntington Library, 1972), 305. For a very moving appreciation of Kipling see, pp. 336-41, 397.

33 "The Endurance Test," Book 5 in Foote, *A Victorian Gentlewoman*, 285-330; and Introduction, 24-7.

34 Wallace Stegner, *Angle of Repose* (New York: Doubleday, 1971).

35 Mary Hallock Foote, "In Exile," *In Exile and Other Stories* (Boston: Houghton, Mifflin, 1894), 1-58, p. 33.

36 Paul Rodman's introduction in *A Victorian Gentlewoman*; Barbara Cragg, "Mary Hallock Foote's Images of the Old West," *Landscape* 24, 3 (1980): 42-7; *Encyclopedia of Frontier and Western Fiction* (New York: McGraw, 1983), s.v., "Foote."

37 Charlotte M. Vaile, *The M.M.C.: a Story of the Great Rockies* (Boston: W. A. Wilde, 1898), 150. Quite possibly, Vaile was inspired by Foote's *In Exile*. Like Frances Newall, the heroine Alice Hildreth is a school teacher in a Colorado mining camp confronted with the world of mining; like Foote the tension between the norms of the East coast es-

tablishment and the West are solved through women's civilizing values; also compare pp. 38, 40. See also Wright *The Winning of Barbara Worth*, Chapter 4; Morrow, *Still Jim*, 271, 340.

38 Elizabeth Foote, "Girl of the Engineers," *Atlantic Monthly* 95 (March 1905): 381-91; L. Frances, "The Engineer," *Overland Monthly* 57, 1 (January 1911): 271-3.

39 William R. Taylor and Christopher Lasch, "Two 'Kindred Spirits': Sorority and Family in New England, 1839-1946," *New England Quarterly* 36 (1963): 25-41; Carroll Smith-Rosenberg, "The Female World of Love and Ritual: Relations between Women in Nineteenth-Century America," *Signs* 1 (Autumn 1975): 1-29; Jean Lipman-Blumen identified homo-social relations as a tool of male power with which women had to compete. "Toward a Homosocial Theory of Sex Roles: An Explanation of Sex Segregation of Social Institutions," *Signs* 1, 3 (Spring 1976) Part II: 15-31, p. 16.

40 John C. Vitale, "The Great Quebec Bridge Disaster," *Consulting Engineer* 38, 2 (February 1967): 92-5.

41 Marjorie Lowry Christie Pickthall, *The Bridge: A Story of the Great Lakes* (New York: Century, 1922).

42 Here, Kipling plays on the word used in India for bridge 'tirtcha' which means both an overpass and a ford in the river, symbolizing pilgrimage. I am grateful to Richard Davis for helping me understand Kipling's story in its British-Indian context.

43 Ann Parry, "Imperialism in 'The Bridge-Builders': Metaphor or Reality?" and "Imperialism in 'The Bridge-Builders': The Builders of the Bridge and the Future of the Raj," *Kipling Journal* 60 (March 1986): 12-22 and (June 1986): 9-16. See also: Michael Adas, *Machines as the Measure of Men: Science, Technology, and Ideologies of Western Dominance* (Ithaca: Cornell University Press, 1989): 235-6; Ammons, "The Engineer as Culture Hero."

44 For example, see the exchange on another bridge disaster, Nathaniel to Anna Ray, September 21, 1916, Nathaniel Chapin Ray Papers, 1878-1916, New Haven Colony Historical Society, New Haven, CT. No critical work or biographical information has appeared on Ray, but see: A.A.C., "In Memoriam: Anna Chapin Ray," *Smith Alumnae Quarterly* 37, 2 (February 1946): 87. *Sheffield Scientific School* (New Haven, 1917), 742-3.

45 Anna Chapin Ray, *The Bridge Builders* (Boston: Little, Brown, 1909), 25, 78, 116, 246

46 Ray, *The Brige Builders*, 198, 392, 397.

47 As cited in Ammons, "Engineer as Cultural Hero," 748-9; Willa Cather, *Alexander's Bridge* (New York: Signet, 1987 [1912]), Intro. by Sharon O'Brien; Tichi, *Shifting Gears*, 173-80; Ziolkowski, "Existential Anxieties."

48 Cather, *Alexander's Bridge*, Introduction; James Woodress in *DLB* 9, s.v., "Cather."

49 Sharon O'Brien, "Becoming Noncanonical," *American Quarterly* 40, 1 (March, 1988); Gilbert and Gubar, *No Man's Land*.

50 Barbara Zabel, "The Machine as Metaphor, Model, and Microcosm: Technology in American Art, 1915-1930," *Arts Magazine* 57 (December 12, 1982): 100-5.

51 Protesting against the decontextualized artistic representation of technology were Diego Riviera's Detroit mural of the Ford factory at River Rouge of 1932 and Lewis Hine's photo-documentary suggestively called *Men at Work*. "This is a book of Men at Work; men of courage, skill, daring and imagination," Hine wrote. "Cities do not build themselves, machines cannot make machines, unless back of them all are the

brains and toil of men. We call this the Machine Age. But the more machines we use the more do we need real men to make and direct them." Lewis W. Hine, *Men at Work* (New York: Dover, 1977 [1932]): Introduction; Ruth Oldenziel, "Artistics Crossings: The Ford (plant) at River Rouge, 1927-1939," in *High Brow Meets Low Brow: American Culture as an Intellectual Concern*, ed. Rob Kroes (Amsterdam: Free University Press, 1988), 37-60.

52 Tichi, *Shifting Gears*. Tichi calls it the gear-and-girder understanding of technology. Competing with the aesthetization of industrial life or what John Stilgoe has coined the "Pullman-car vision of the industrial zone" were the depictions in word and image of Progressive reformers concerned with the working and living conditions of immigrants in back alleys of the industrial cities. John R. Stilgoe, "Molding the Industrial Zone Aesthetic, 1880-1929," *Journal of American Studies* 16 (April 1982): 5-24, p. 12. This understanding was not incorporated into the definition of technology, however.

53 Molly Nesbitt, "Ready-Made Originals: the Duchamp Model," *October* 37 (1986): 53-64. For the Russian Constructivist movement, see L. Lissitzky's drawing experiments exploring the differences and similarities between art and engineering, and Varvara Stepanova and Aleksandr Rodchenko's development of working methods and curriculum. *Art into Life: Russian Constructivism 1914-1932*, Minneapolis, Walker Art Center, October 7 - December 30, 1990. At the Bauhaus, Walter Gropius led the art school into a firm alliance with industry, against the proponents of the crafts such as Johannes Itten, a direction Hannes Meyer appropriately called a polytechnic education. Frank Whitford, *Bauhaus* (London: Thames and Hudson, 1984), Chapter 11, and documents 202-10; Krzysztof Fijalkowski, "Dada and the Machine," *Journal of European Studies* 17 (1987): 233-51.

54 On the elusive but fascinating Stevens: Francis Naumann, "A Lost American Futurist," *Art in America* (April 1994); Carolyn Burke and Naomi Sawelson-Gorse, "In Search of Frances Simpson Stevens," *Art in America* (April 1994); Carolyn Burke, *Becoming Modern. The Life of Mina Loy* (Berkeley: University of California Press, 1996); Naumann personal files; and Frances Simpson Stevens materials in possession of the author.

55 *291* nos. 5-6 (New York, July-August 1915). The New York avant-garde journal, established by Stieglitz with financial backing from Haviland, became a fertile meeting ground for artist-refugees from Europe and their American colleagues for formulations on modernism during the war. On Loy's attempts to patent her inventions, see, "Invention Corselet, n.d." Series II, Box 6, File 186, Mina Loy Papers, Beinecke Rare Book and Manuscript Library, Yale University, New Haven, CT. and Burke, *Becoming Modern*.

56 Ernest Gombrich, "Visual Metaphors of Value in Art," *Meditations on a Hobby Horse: And Other Essays on the Theory of Art* (1963), 12-29. On the engineering vision as process in photography, see Terry E. Smith, "Tracing the 'Eye of Power': Foucault, Factories and Photography, U.S.A. 1900-1930" (paper presented at the College Art Association Conference, Washington, February 23, 1991). For a general overview, Richard Wilson, Dianne H. Pilgrim, and Tashjian, *The Machine Age in America, 1918-1941* (New York: Brooklyn Museum, 1986).

57 Paul Haviland, *291* (March 1915).

58 Burke, *Becoming Modern*, 217-8.

59 Burke, *Becoming Modern*, 178-84
60 See for a rich and general discussion on the issues of cross-dressing, transsexuality, passing, camp, and voguing as forms of subversion or surveillance: Anne McClintock, *Imperial Leather: Race, Gender and Sexuality in the Colonial Contest* (New York: Routledge, 1995); Marjoire Garber, *Vested Interests. Cross Dressing and Cultural Authority* (New York: Routledge, 1992).

CHAPTER 5

1   Unless otherwise indicated, what follows is based on my survey of *all* American accredited engineering schools prior to 1945. A query was sent to 274 schools. This correspondence was followed up by letters to registrar's offices, engineering colleges, university archives and alumni associations. In total, 175 schools answered: 82 reported on women graduates. Of the 99 schools that did not respond, 10 appear to have awarded a significant number of degrees to women. Based on this survey and other sources, over 600 women in engineering have been identified by name. *Women Engineering Graduates in the U.S., 1877-1945 Papers*, Amsterdam, the Netherlands, (Women Engineering Graduates Papers, hereafter). Other sources include: The Society of Women Engineers Collection, Walter Reuther Library, Wayne State University, Detroit, MI; The Society of Women Engineers's directories of the 1950s; The Gilbreth Papers at Purdue University and Purdue's Institutional Archives, Lafayette, IL; Institute Archives and Special Collections, MIT Libraries, Cambridge, MA; The Records of the Women's Bureau, Bulletins of the Women's Bureau, RG 86, Boxes 699-701, National Archives; Census Records, *Who's Who in Engineering (1922-23)*; and General Electric Hall of History Collection, Schenectady, NY.
2   Author's interview with Nora Blatch's daughter, Rhoda Barney Jenkins, September 1989; Ellen Dubois, ed., "Spanning Two Centuries: The Autobiography of Nora Stanton Barney," *History Workshop Journal* 22 (1986): 131-52, p. 134; cf. Suzy Fisher, "Nora Stanton Barney, First U.S. Woman CE, Dies at 87," *Civil Engineering* 41 (April 1971): 87.
3   Examples of hostility: Vera Jones Mackay in *100 years: A Story of the First Century of Vanderbilt University School of Engineering, 1875-1975* (Nashville: Vanderbilt Engineering Alumni Association, 1975), 116-9; Marion Monet (MIT '43) personal interview March 23, 1990; Author's telephone interview with Dorothy Quiggle, November 14, 1989 and *MIT Survey*; example of aloofness: Eleonore D. Allen (Swarthmore '36), "Lady Auto Engineer: Her Ideas Irreparable," *New York World-Telegram & Sun* (August 14, 1961). On the post-suffrage generation of women professionals, see: Nancy Cott, *The Grounding of Feminism* (New Haven: Yale University Press, 1987): Chapter 1 and Introduction.
4   Rosalyn Rosenberg, *Beyond Separate Spheres. Intellectual Roots of Modern Feminism* (New Haven: Yale University, 1983); Louise Michele Newman, ed., *Men's Ideas/Women's Realities* (New York: Pergamon Press, 1985); Ruth Oldenziel, Karin Zachmann, and Annie Canel, *Crossing Boundaries: The History of Women Engineers in the Comparative Perspective* (London: Harwood Academic Press, Forthcoming). John

M. Staudenmaier, S.J., *Technology's Storytellers. Reweaving the Human Fabric* (Cambridge: MIT Press, 1985); Margaret W. Rossiter, *Women Scientists in America. Struggles and Strategies to 1941* (Baltimore: The Hopkins University Press, 1982).

5 Lillian Gilbreth, "Marriage, a Career and the Curriculum," typewritten manuscript (probably 1930s), Lillian Gilbreth Collection, NHZ 0830-27, Box 135, Department of Special Collections and Archives, Purdue University Library, Lafayette, IL. (Gilbreth papers, hereafter). See also, "American Women Survey Their Emancipation. Careers Are Found an Aid To Successful Married Life," *Washington Post* (August 16, 1934); Edna Yost, *Frank and Lillian Gilbreth, Partners for Life* (New Brunswick: Rutgers University Press, 1949) and Yost's description in *American Women of Science* (New York: Frederick A. Stokes, 1943); Ruth Schwartz Cowan, *Dictionary of Notable Women* (1980), s.v., "Gilbreth." The notion of "borrowed identity" comes from Margot Fuchs, "Like Fathers-Like Daughters: Professionalization Strategies of Women Students and Engineers in Germany, 1890s to 1940s," *History and Technology* 14, 1 (1997): 49-64.

6 See Rossiter's excellent *Women Scientists in America* (1982), 248. The sequel *and her Women Scientists in America. Before Affirmative Action, 1940-1972* (Baltimore: The Hopkins University Press, 1995) is equally path breaking and continues to be an inspiration.

7 *First Century of Vanderbilt University School of Engineering*, 116-9; Correspondence, Vanderbilt University, Special Collections University Archives, Nashville, TN., (Women Engineering Graduates Papers).

8 Ovid Eshbach, Technological Institute, Northwestern University, Chicago, IL, February 15, 1945 with Woman's Bureau, RG 86, Box 701, Woman's Bureau, National Archives, Washington, DC. [WB NA, hereafter]. Cf. Juliane Mikoletzky, "An Unintended Consequence: Women's Entry into Engineering Education in Austria," *History and Technology* 14, 1 (January 1997): 31-48. Fuchs, "Like Fathers-Like Daughters."

9 Surveying many unusual fields of women's employment, Miriam Simons Leuck reported on a great many widows, some of whom were engineers: "Women in Odd and Unusual Fields of Work," *AAAPPS* 143 (1929): 166-79. U.S. Bureau of the Census, *Census* (Washington, DC: Government Printing Office, 1890).

10 A., "Women Engineers," Letter to the Editor, *Professional Engineer* (May 1922): 20; "Roebling Memorabilia," *The New York Times* (October 4, 1983); Gustave Lindenthal, Letter to the Editor, "A Woman's Share in the Brooklyn Bridge," *Civil Engineering* 3, 3 (1933): 473; Alva T. Matthews, "Emily W. Roebling, One of the Builders of the Bridge," in *Bridge to the Future: A Centennial Celebration of the Brooklyn Bridge*, eds. Margaret Latimer, Brooke Hindle, and Melvin Kranzberg *Annals of the New York Academy of Sciences* (1984), 63-70; for a biography on Roebling, Marilyn Weigold's study, see *The Silent Builder: Emily Warren Roebling and the Brooklyn Bridge* (Port Washington, NY: Associated Faculty Press, 1984). For an appreciation of her work by the engineering community, see "Engineers Pay Tribute to the Woman Who Helped Build the Brooklyn Bridge" (address by David B. Steinman for the Brooklyn Engineers Club, May 24, 1953), reprinted in *The Transit of Chi Epsilon* (Spring-Fall 1954): 1-7.

11 For a treatment of Gilbreth's work, see: Martha Moore Trescott, "Lillian Moller Gilbreth and the Founding of Modern Industrial Engineering," in *Machina Ex Dea:*

Feminist Perspectives on Technology, ed. by Joan Rothschild (New York: Pergamon Press, 1983), 23-37. "Marriage, a Career and the Curriculum"; "American Women Survey Their Emancipation"; Yost, *Frank and Lillian Gilbreth* and *American Women of Science*. Yost was one of the Gilbreths' most important promoters and popularizers. Ruth Schwartz Cowan in *Notable American Women* (1980), s.v., "Gilbreth."

12 Correspondence Special Collections Department, University Archives, University of Nevada, Reno, correspondence with New Jersey Institute of Technology, Alumni Association, Newark, NJ, (Women Engineering Graduates Papers); Carolyn Cummings Perrucci, "Engineering and the Class Structure," in *The Engineers and The Social System*, eds. Robert Perrucci and Joel Gerstl (New York: John Wiley and Sons, 1969), 284, Table 3.

13 Correspondence, Cornell University Library, Department of Manuscripts and University Archives, (Women Engineering Graduates Papers); Eve Chappell, "Kate Gleason's Careers," *Woman's Citizen* (January 1926): 19-20, 37-8; *DAB*, s.v., "Gleason"; "A Woman Who Was First," *The Cornell Alumni News* (January 19, 1933): 179 with excerpts from *The Cleveland Plain Dealer* and *The New York Tribune*; Leuck, "Women in Odd and Unusual Fields," 175; *The Gleason Works, 1865-1950* (n.p., 1950); *ASME. Transactions* 56, RI 19 (1934), s.v., "Gleason." Cf. Christopher Lindley in *Notable American Women* (1934), s.v., "Gleason."

14 Judith S. McIlwee and Gregg J. Robinson, *Women in Engineering: Gender, Power and Workplace Culture* (Albany: State University of New York Press, 1992).

15 DuBois, "Spanning Two Centuries," 150.

16 Terry Kay Rockefeller *Notable American Women* (1980), s.v., "Barney."

17 "Barnard Girls Test Wireless' Phones," *The New York Times* (February 23, 1909): 7:3.

18 In *Inventing American Broadcasting, 1899-1922* (Baltimore: The Johns Hopkins University, 1987), 167-7, Susan J. Douglas writes that Nora Blatch's "contributions to early development of voice transmission have been either completely ignored or dismissed", (p. 174).

19 "Warns Wives of 'Careers,'" *The New York Times* (July 28, 1911) 18:3 and response by Ethel C. Avery, "Suffrage Leaders and Divorce," *The New York Times* (July 31, 1911) 6:5. See also Margaret W. Raven, a graduate of MIT ('39) in General Science and Meteorology, Association of MIT Alumnae, Membership Survey, 1972, MIT, Institute Archives, Cambridge, MA; and McIlwee and Robinson, *Women in Engineering*.

20 Edna May Turner, "Education of Women for Engineering in the United States, 1885-1952," (Ph.D. diss., New York University, 1954). This valuable and pioneering dissertation contains little analysis or biographical information beyond the statistics. Cf. Annie Canel, "Following the Trail of the Pioneers: Women Engineers in the French 'Grand Ecoles' in the 1960s and 1970s," *History and Technology* 14 (1997): 123-45.

21 Correspondence with University Archives, Ohio State University, Ames, OH; Special Collections and Center for Southern History and Culture, University of Alabama, Tuscaloosa, AL; correspondence with Stanford Alumni Association, Stanford, CA, (Women Engineering Graduates Papers). Robin found a similar trend for the post-World War era. "The Female in Engineering," in *The Engineers and the Social System*, 203-18. Cf. "Report of the Committee on Statistics of Engineering Education,"

Proceedings of the Society for the Promotion of Engineering Education 10 (1902): 230-57, p. 238, Table I.

22 Tom S. Gillis, roommate of Henry M. Rollins, Hutson's son, letter to author, October 21, 1989; correspondence, Texas A&M University, College Station, TX; Library and Learning Resources, Univeristy of Missouri-Rolla, Rolla, MO, (Women Engineering Graduates Papers); Lawrence O. Christensen and Jack B. Ridley, *UM-Rollo: A History of MSM/UMR* (Missouri: University of Missouri Printing Services, n.d.), 93-4; and his "Being Special: Women Students at the Missouri School of Mines and Metallurgy," *Missouri Historical Review* 83, 1 (October, 1988): 17-35; Frances A. Groff, "A Mistress of Mechanigraphics," *Sunset Magazine* (October 1911): 415-8.

23 Clipping file, Alumni Office, Swarthmore College, Swarthmore, PA; for further information: *Yearbook* (1942), (Women Engineering Graduates Papers); Edward M. Tuft, "Women in Electronics," *National Business Women* (November 1956). See also: Olive W. Dennis to Marguerite Zapoleon, September 3, 1947, Women's Bureau, Bulletins, RG 223 no. 5, WB NA.

24 Sally Hacker, "The Mathematization of Engineering: Limits on Women and the Field," in *Machina Ex Dea*, 38-58. After the second World War, competence in mathematics was the single most common denominator in women's motivation to go into engineering. Martha Moore Trescott, "Women Engineers in History: Profiles in Holism and Persistence," in *Women in Scientific and Engineering Professions*, eds. Violet B. Haas and Carolyn C. Perrucci (Ann Arbor: The University of Michigan Press, 1984), 181-205. More work needs to be done on women's education in mathematics at the secondary school level, but see Warren Colburn, "Teaching of Arithmetic. Address before the American Institute of Instruction in Boston, August 1830," in *Readings in the History of Mathematics Education* (Washington, DC: National Council of Teachers in Mathematics, 1970), 24-37; Ruth Oldenziel, "The Classmates of Lizzie Borden" (University of Massachusetts at Amherst: unpublished manuscript, 1982). See also Robert Fox and Anna Guagnini, "Classical Values and Useful Knowledge: The Problem of Access to Technical Careers in Modern Europe," *Daedalus* 116, 4 (1987): 153-71. For an excellent discussion on the issue during the 1970s and 1980s, see McIlwee and Robinson, *Women in Engineering*

25 Alice C. Goff, *Women Can Be Engineers* (Youngstown, OH: Privately Printed, 1946); "Edith Clarke-Biographical Data," typewritten," General Electric Hall of History, Schenectady, NY [GE History Hall, hereafter; "Miss Clark Dies at 76: Retired Engineer at GE," *Schenectady Gazette* (November 19, 1959); "Teaching Opens New World to Woman, 65," *Dallas Morning News* (December 12, 1948); *Who's Who* 40 (1978-79), s.v., "Eaves"; College of Engineering and Applied Science, Special Collections, University of Colorado, Boulder, CO; correspondence with College of Engineering, Office of Development, Bentley Historical Library, University of Michigan, Ann Arbor, MI, (Women Engineering Graduates Papers).

26 Nora Blatch Photo Album. Courtesy of her daughter Rhoda Barney Jenkins, Greenwich, CT. (Barney-Jenkins Papers, hereafter); Records of the Women's Bureau, Women's Bureau Bulletin 223 no. 5; RG, Box 701, WB NA. "Rice Women Engineering," *Rice Engineer* (January 1986): 18-23. *Women Engineers and Architects* (March 1938): 1, Box folder "Women Engineers and Architects, 1938-1940," Society of Women

Engineers Collection, Walter Reuther Library, Wayne State University, Detroit, MI (SWE Collection, hereafter); correspondence, Woodson Research Center, Rice University, Houston, TX.

27 Karl Drews, "Women Engineers: The Obstacles in Their Way," *Scientific American, Supplement* 65 (March 7, 1908): 147-8.

28 Barbara Drygulski Wright's in her introduction to *Women, Work, and Technology. Transformations* (Ann Arbor: University of Michigan Press, 1987), 16-7.

29 John W. Upp, "The American Woman Worker," *The Woman Engineer* 1 and "American Women Engineers," *The Woman Engineer* 1, 11 (June 1922): 156; 186-88.

30 *Report of the Registrar of the University of Michigan, 1926-1941*; correspondence, Office of Development, Bentley Historical Library, University of Michigan, Ann Arbor, MI, (Women Engineering Graduates Papers).

31 College of Engineering and Applied Science; Special Collections, University of Colorado, Boulder, CO; correspondence University of Nevada, Special Collections Department, Reno, NV; Ohio Northern University, Alumni Office, Ada, OH; *Alumni Directory* (1875-1953), University of Minnesota; *The Minnesota Techno-Log* (May 1925): 11; correspondence University of Minnesota, University Archives, Minneapolis, MN, (Women Engineering Graduates Papers).

32 correspondence with Alumni Records Office, College of Engineering, Bentley Historical Library, University of Michigan, Ann Arbor, MI; Correspondence with Bertha L. Ihnat, Ohio State University; *Commencement Programs* (1878-1907), Ohio State University, Ames, OH, (Women Engineering Graduates Papers); Clipping file, General Electric Company, GE History Hall; Goff, *Women Can Be Engineers*; Elsie Eaves, "Wanted: Women Engineers," *Independent Woman* (May 1942): 132-3, 158-9, p. 158.

33 Correspondence with Alumni Association, Iowa State University, Ames, IA, (Women Engineering Graduates Papers); Adelaide Handy, "Calculates Power Transmission for General Electric Company," *The New York Times* (October 27, 1940).

34 Dennis to Marguerite Zapoleon, September 3, 1947, WB NA.

35 Nora Stanton Barney letters to the editor, "Industrial Equality for Women," *NY Herald-Tribune* (April 21, 1933) and "Wages and Sex," *N.Y. Herald-Tribune* (July 21, 1933).

36 Elsie Eaves to Mary Esther Poorman, November 8, 1933; Elsie Eaves to Virginia A. Swaty, March 2, 1936; Elsie Eaves to Jane Hall, November 10, 1938, Box 146, folder "Earliest Efforts to organize, 1920-1940," SWE Collection.

37 Box 146, folder "Earliest efforts to organize, 1920-1940," SWE Collection.

38 Juliet K. Coyle, "Evolution of a Species – Engineering Aide," *U.S. Woman Engineer* (April 1984): 23-4; Robert McMath, Jr. et al., *Engineering the New South. Georgia Tech, 1885-1985* (Athens: The University of Georgia Press, 1985), 212; correspondence, John D. Akerman with Curtiss-Wright Corporation, 1942-43, University Archives, University of Minnesota, Minneapolis, MN, (Women Engineering Graduates Papers); Curtiss-Wright Engineering Cadettes Program Papers, Archives of Women in Science and Engineering, Iowa State University, Ames, IA; correspondence George Institute of Technology, Archives and Records Department, Atlanta, GA; "Engineering Aide, Curtis Wright Program Follow-up of Cadette trained at Rennselaer Polytechnic," March 22, 1945, Rennselaer Polytechnic Library, NY, (Women Engineering Graduates Papers).

39 C. Wilson Cole, "Training of Women in Engineering," *Journal of Engineering Education* 43 (October 1943): 167-84; E.D. Howe, "Training Women for Engineering Tasks," *Mechanical Engineering* 65 (October 1943): 742-4; R. H. Baker and Mary L. Reimold, "What Can Be Done to Train Women for Jobs in Engineering," *Mechanical Engineering* 64 (December 1942): 853-5; D. J. Bolanovich, "Selection of Female Engineering Trainees," *Journal of Educational Psychology*: 545-53; "Free-Tuition in Courses Engineering for Women," *Science* 95, 2455, Suppl. 10 (January 9, 1942): 10; "Training for Women in Aeronautical Engineering at the University of Cincinnati," *Science* 97 (June 18, 1943): 548-9.

40 *Training in Business and Technical Careers for Women* (Ohio: The University of Cincinnati, 1944); Harriette Burr, "Guidance for Girls in Mathematics," *The Mathematics Teacher* 36 (May 1943): 203-11. Margaret Barnard Pickel, "A Warning to the Career Women," *The New York Times Magazine* (July 16, 1944): 19, 32-3 and Malvina Lindsay, "The Gentler Sex. Young Women in a Hurry," *Washington Post* (July 20, 1944).

41 Olive W. Dennis to Marguerite Zapoleon, September 3, 1947, WB NA.

42 Many examples may be found in the Records of the Women's Bureau Bulletins, RG 86, WB NA.

43 Coyle, "Evolution of a Species"; U.S. Department of Labor, Woman's Bureau, *The Outlook for Women in Architecture and Engineering*, Bulletin 223 no. 5 (Washington, DC: Government Printing Office, 1948); U.S. Department of Labor, Woman's Bureau, "Employment and Characteristics of Women Engineers," *Monthly Labor Review* (May 1956): 551-6.

44 Excerpts from Olive W. Dennis, "So – Your Daughter Wants to be a Civil Engineer," Box 701, WB NA; excerpts from letter, *Baltimore and Ohio Magazine* (September 1940): 30.

45 Olive W. Dennis, Clipping file, Baltimore and Ohio Railroad Museum, Baltimore, MD.

46 Editorial, "Women AIME Members Contribute Their Share in Engineering War," *Mining and Metallurgy* 23 (November 1942): 580-1.

47 For gender differences in the professions: Joan Brumberg and Nancy Tomes, "Women in the Professions: A Research Agenda for American Historians," *Reviews of American History* 10, 2 (June 1982): 275-96; Barbara F. Reskin and Polly A. Phipps, "Women in Male-Dominated Professional and Managerial Occupations," in *Women Working: Theories and Facts in Perspective*, eds. Ann H. Stromberg and Shirley Harkness (Mountain View, CA: Mayfield, 1988); Barbara F. Reskin and Partricia A. Roos, *Job Queues, Gender Queues: Explaining Women's Inroads into Male Occupations* (Philadelphia: Temple University Press, 1990). See also, Andrew Abbott, *The Systems of Profession: An Essay on the Division of Expert Labor* (Chicago: Chicago University Press, 1988), 98-111.

48 The number of women in the professional organizations was reported in Woman's Bureau, *The Outlook for Women*, 22. In 1946, women accounted for six out of every thousand members in the ASCE and AIEE. Nine participated in the AIChE, 16 in the ASME, and 21 in the AIME. These figures roughly resemble the data gathered in the survey. New research should focus on local engineering organizations, however.

49 A. Michal McMahon, *The Making of a Profession: A Century of Electrical Engineering in America* (New York: The Institute of Electrical and Electronics Engineers Press, 1984), 58-9.

50 Anson Marston to Hilda Counts, May 6, 1919, Box 146, folder "Earliest Efforts to organize, 1918-1920," SWE Collection. On her suit and efforts to rally support for her case, see W. W. Pearse to Nora S. Blatch, January 20, 1915 and another from Ernest W. Schroder to Nora S. Blatch, January 20, 1915, Barney-Jenkins Papers. "Mrs. De Forest Loses Suit," *The New York Times* (January 22, 1916): 13; Reports on the case appeared also on Saturday, January 1, 1916: 18, "Mrs. De Forest Files Suit"; January 12, 1916: 7; "Old Men Bar Miss Blatch"; and in *The New York Sun* January 1, 1916: 7, "Mrs. De Forest, Suing, Tells her Real Age." Blatch's employment history may be found in *Notable American Women* (1980), s.v., "Barney"; DuBois, "Spanning Two Centuries," 148; Speech "Petticoats and Slide Rules," p. 6, Box 187, folder "Miscellaneous Correspondence, Elsie Eaves," SWE Collection. The speech was published under the same title in a slightly altered form in *The Midwest Engineer* (1952).

51 "Women Engineers-Yesterday and Today," *The Bent of Tau Beta Pi* (Summer 1971): 10-2; correspondence, University Library, University Achives, University of Illinois at Urbana-Champaign, Urbana, IL, (Women Engineering Graduates Papers).

52 Rossiter, *Women Scientists in America* (1982), 77 and Chapter 4. For a sample of the literature on women in the professions see Barbara Melosh, *The Physicians' Hand: Work Culture and Conflict in American Nursing* (Philadelphia: Temple University Press, 1982); Brumberg and Tomes, "Women in the Professions."

53 *Dictionary of Occupational Titles* (1939); Arthur D. Little to Beatrice Doerschuk, February 6, 1922, Bureau of Vocational Information, Schlesinger Library, Radcliffe College, Cambridge, MA, microfilm [BVI hereafter], reel 12. His opinion was extensively quoted in a section on opportunities in chemical engineering in the Bureau of Vocational Information, *Women in Chemistry: A Study of Professional Opportunities* (New York: Bureau of Vocational Information, 1922), 60-1. Terry S. Reynolds, "Defining Professional Boundaries: Chemical Engineering in the Early 20th Century," *Technology and Culture* 27 (1986): 694-716, p. 709. On Little, see also David F. Noble, *America By Design* (New York: Oxford University Press, 1977), 124-5.

54 Calco Chemical Company, M.L. Crossley to Emma P. Hirth, December 24, 1919; National Aniline & Chemical Company, C. G. Denck to Emma P. Hirth, December 22, 1917; A. P. Tanberg, Dupont Co. to Emma P. Hirth, August 30, 1921; L.C. Drefahl Grasselli Chemical Company to Emma P. Hirth, December 20, 1917; all BVI reels 12 and 13.

55 Mrs. Glenola Behling Rose, chemist at Dupont Company, questionnaire, February 1920, BVI reel 12; Florence Renick, questionnaire, February 1920, BVI reel 13; Jessie Elizabeth Minor, Questionnaire, January 12, BVI reel 12.

56 U.S. Department of Labor, Women's Bureau, *The Occupational Progress of Women, 1910 to 1930*, Bulletin 104 (Washington, DC: Government Printing Office, 1933); Woman's Bureau, *The Outlook for Women*, 46; (Women Engineering Graduates Papers).

57 U.S. Department of Interior, Office of Education, *Land-Grant Colleges and Universities* (Washington, DC: Printing Office, 1930), 805; Ruth Oldenziel, "Gender and the

Meanings of Technology: Engineering in the U.S., 1880-1945" (Ph.D. diss., Yale University, 1992), Fig. 6.

58   David M. Blank and George J. Stigler, *The Demand and Supply of Scientific Personnel*, General Series, 62 (New York: National Bureau of Economic Research, 1957), 4, 8-9, 10-2, 87, 192.

59   For example, author's interview with Marion Monet (MIT '43), March 23, 1990. Author's telephone interview with Dorothy Quiggle, November 14, 1989 and *MIT Survey*; Cott, *The Grounding of Feminism*, Chapter 1 and Introduction.

60   Box 146, folder "Earliest Efforts to Organize, 1918-1920," SWE Collection.

61   Carroll W. Pursell, "'Am I a Lady or an Engineer?' The Origins of the Women's Engineering Society in Britain, 1918-1940," *Technology and Culture* 34, 1 (January 1993): 78-97. See also: Crystal Eastman, "Caroline Haslett and the Women Engineers," *Equal Rights* 11/12 (10 April 1929): 69-70.

62   "Origins of the Society by Phyllis Evans Miller," Box 147, SWE Collection; "Girls Studying Engineering See Future for Women in These Fields," *The Christian Science Monitor* (Saturday, April 16, 1949): 4; "New Members of the Women's Engineering Society" *Women's Engineering Society* (1950), 315.

63   "Girls Studying Engineering See Future"; Mart Navia Kindya with Cynthia Knox Lang, *Four Decades of the Society of Women Engineers* (Society of Women Engineers, n.d.), 12; Lillian G. Murad to Katherine (Stinson), June 8, 1952, Box 146, folder "SWE history 1951-1957," SWE Collection.

64   See also Carroll W. Pursell, *The Machine in America: A Social History of Technology* (Baltimore: The Johns Hopkins University Press, 1995), 310 for a passing remark on this issue.

65   Rossiter, *Women Scientists in America* (1995), 28, 59, and Chapter 2.

66   Lillian M. Gilbreth, "The Daughters of Martha," speech before the Society of Women Engineers at the opening of the headquarters in the United Engineering Building, SWE celebration banquet, 1961. Box 24, Lillian Gilbreth Papers.

67   Today, the SWE's office is overcrowded and understaffed, occupying a tiny space in an otherwise imposing building where all the engineering societies reside together near the United Nations Headquarters in New York city.

68   "Marriage, a Career and the Curriculum," Box 135, Gilbreth Papers.

69   For a provocative inquiry, see: Pamela Mack "What Difference has Feminism Made to Engineering in the 20th Century?" in *Science, Medicine, Technology: The Difference Feminism Has Made*, eds. Londa Schiebinger, Elizabeth Lunbeck, and Angela N.H. Creager (Chicago: University of Chicago Press, Forthcoming).

70   Mary Ritter Beard, *Woman's Work in Municipalities* (New York: Arno Press, 1972 [1915]).

Epilogue

1   *The New Encyclopaedia Britannica* (Chicago: Encyclopaedia Britannica, 1974-1985), 15th ed., s.v., "Technology," 452; Richard M. Rollins, "Words as Social Control: Noah Webster and the Creation of the *American Dictionary*," *American Quarterly* 28 (1976):

415-30; see also Eva Mae Burkett, *American Dictionaries of the English Language Before 1861* (Metuchen, NY: Scarecrow Press, 1979).

2   William A. Lydgate, "The 32 Million-Work Classic," *Saturday Review* 25, 18 (May 2, 1942): 3-4, 22-3. Willard Huntington Wright, *Misinforming a Nation* (New York: B. W. Huebsch, 1917), 19, contested *Britannica*'s canonical claims. Cf. in his review, "The Engineer and the Encyclopaedia," *Cassier's* 40 (December, 1911): 753-60, Herry Harrison presented engineers as scholars rather than as shopfloor-trained technicians.

3   Ann J. Lane, *Mary Ritter Beard: A Sourcebook* (New York: Schocken Books, 1977), 47, see also Chapters 3 and 5. For Gage's influence, see Beard's letter to Mrs. Schlesinger, where she writes that "[o]nly Matilda Joslyn Gage seems to have known much about the history of women and she naturally saw it as a history of subjection to sadism via Canon Law in which she rather specialized," Beard to Mrs. Schlesinger, June 15, 1944, in *A Woman Making History. Mary Ritter Beard Through Her Letters*, ed. and introd. by Nancy F. Cott (New Haven: Yale University Press, 1991), 248. For indications of the influence of evolutionary anthropologists's theories on Beard's thinking and Mason in particular, see in addition the assessment of the entry of *Social Anthropology*, Lane, 219-20; Cott, *A Woman Making History*, 26-27, 32. For an early letter on her understanding of women's inventions, see: Mary Beard, "Inventions Are for Men," Letter to the Editor, *The New York Times* (August 17, 1915). See also, Mary Ritter Beard, *Woman as Force in History* (New York: Macmillan, 1946).

4   Cott, *A Woman Making History*, 51, 221, 227, 237-8, 247-51, 283-4.

5   *Encyclopaedia Britannica*, 14th ed. (1968), s.v., "History of Technology," by Robert S. Woodbury, Professor of the History of Technology at MIT since 1963.

6   Telephone interview with Melvin Kranzberg, September 15, 1990. Prior to his consultant work for the *Encyclopaedia Britannica*, he wrote its entries on: "Industrial Revolution" (1964) and "Inventions and discoveries" (1967). His encyclopedic oeuvre did not stop here: he wrote for *Cowles Comprehensive Encyclopedia* "Transportation" (1967), for *An Encyclopedia of World History* sections on technological developments (1968); for the *World Book Encyclopedia* the following entries: "Invention" and "Technology" (1972), "Transportation" (1981 and subsequent editions). Kranzberg was also a special consultant for the *Harper Encyclopedia of Science* in 1962, a consultant for the *Cowles Comprehensive Encyclopedia* from 1963 to 1967, an adviser for the *The New Columbia Encyclopedia* from 1974 to 1977. Melvin Kranzberg Papers, Archives Center, National Museum of American History, Smithsonian Institution, Washington, DC, (Kranzberg Papers, hereafter).

7   Philip W. Goetz, Executive Editor, The Encyclopaedia Britannica, to Eugene S. Ferguson, May 9, 1967, Series I, Box 42, Kranzberg Papers.

8   Mortimer J. Adler, "The Circle of Learning," *Encyclopaedia Britannica, Propaedia: Outline of Knowledge* (Chicago: The Encyclopaedia Britannica, 1974), 5-9, p. 6.

9   For example, they stressed how the scientific observations by Sumerian astrologists were used to benefit their country's irrigation systems and how the theories of Boyle, Guericke, and Papin had paved the way for the invention of the steam engine (462). Bruce E. Seeley, "SHOT, the History of Technology and Engineering Education," *Technology and Culture* 36, 4 (October 1995): 739-72.

10 The New Encyclopaedia Britannica, *Propaedia: Outline of Knowledge*, 434-97.

11 Eugene S. Ferguson to Warren E. Preece, Editor-in-Chief, The Encyclopaedia Britannica, June 23, 1967, Box 42, Series I, Kranzberg Papers. Franz Reuleaux, *Theoretische Kinematik: Grundzüge einer Theorie des Maschinenwesens* (Braunschweig, 1875) translated as "Technology and Civilization," *Annual Report of the Smithsonian Institution: General Appendix* (Washington, DC: Government Printing Office, 1890), 705-20. See also: Jeffrey Herf, "The Engineer as Ideologue: Reactionary Modernists in Weimar and Nazi Germany," *Journal of Contemporary History* 19 (1984): 631-48; Karin Zachmann, "'Involvement with Technology Does not Harm Their Charm and Femininity'. Mobilization, Modernization, and Ambivalence of Women Engineers in East Germany," *Crossing Boundaries: History of Women Engineers in Cross-Cultural Comparison,* eds. Ruth Oldenziel, Karin Zachmann, and Annie Canel (London: Harwood Academic Press, Forthcoming).

12 Eugene S. Ferguson, "Kinematics of Mechanisms from the Time of Watt," in *Contributions from the Museum of History and Technology* 27 (1963), 186-230, pp. 216-27. French philosopher and theologian Jacques Ellul (b. 1912) undermined the treatment of technology as an artifact with which Ferguson and Kranzberg wrestled. Jacques Ellul, *La Technique* (1954), published in English under the title *The Technological Society* (1963), defined *"technique"* as "the totality of methods, rationally arrived at and having absolute efficiency." The Center for the Study of Democratic Institutions, responsible for disseminating of Ellul's work to an American audience and for organizing a conference in 1962 under the auspices of the *Encyclopaedia Britannica* called *"The Technological Order,"* homage to Ellul's work. The conference would become the basis for the *Britannica's* entry on Technology providing the newly established Society of the History of Technology and its journal *Technology and Culture* the financial and intellectual boost they desperately needed. In preparation for the conference on Technology at the Center, Robert M. Hutchins and Scott Buchanan sought advice for a European contribution from Aldous Huxley, who recommended Ellul. Merton and Buchanan endorsed the translation of the book ito English. Jacques Ellul, *The Technological Society* (New York: Knopf, 1963): Statement from the publisher. Telephone interview with Melvin Kranzberg, August 31, 1990. See, for the proceedings of the Conference, the special issue of *Technology and Culture* (1963). Robert M. Hutchins was President and chairman of the Board of Editors of the *Britannica* from 1947 until 1974 and president of the Center for the Study of Democratic Institutions from 1954 until 1974. He had also been president of the University of Chicago between 1929 and 1945, as well as its chancellor from 1945 to 1953. *Proceedings of the Encyclopedia Britannica Conference on the Technological Order,* ed. Carl F. Stover (Detroit: Wayne State University Press for the Society of the History of Technology, 1962).

13 For a treatment of the historiographical debate, see John M. Staudenmaier, S.J., *Technology's Storytellers: Reweaving the Human Fabric* (Cambridge: MIT Press, 1985); Mark H. Rose, "Science as an Idiom in the Domain of Technology," *Science and Technology Studies* 5, 1 (1987): 3-11; Arthur P. Molella and Nathan Reingold, "Theorists and Ingenious Mechanics: Joseph Henry Defines Science, *Science Studies* 3, (1973): 322-35; Ronald Kline, "Construing 'Technology' as 'Applied Science'. Public Rhetoric of Scientists and Engineers in the United States, 1880-1945," *ISIS* 86 (1995): 194-221. For example, the engineer Vannevar Bush entitled his blueprint for government support for applied

research, *Science, The Endless Frontier: A Report to the President* (Washington, DC: Government Printing Office, 1945); Arthur P. Molella, "The Museum That Might Have Been: The Smithsonian's National Museum of Engineering and Industry, Technology and Culture," *Technology and Culture* 32, 2, pt. I, (April, 1991): 237-63.

14 This approach is most obvious in "Part Seven. Technology," *Propaedia: Outline of Knowledge*, 265-97.

15 Seeley, "SHOT, the History of Technology and Engineering Education"; H. W. Lintsen and E. Homburg, "Techniekgeschiedenis in Nederland," *Geschiedenis van de Techniek in Nederland. De wording van een moderne samenleving, 1800-1890* IV (Zutphen: Walburg Pers, 1995), 255-66.

16 Jan Zimmerman, *Once Upon the Future: A Woman's Guide to Tomorrow's Technology* (London: Pandora Press, 1986), Introduction; Martha Moore Trescott, ed., *Dynamos and Virgins Revisited: Women and Technological Change in History* (Metuchen, NJ: Scarecrow Press, 1979); Joan Rothschild ed., *Machina Ex Dea: Feminist Perspectives on Technology* (New York: Pergamon Press, 1983); Judith A. McGaw, "Review Essay: Women and the History of American Technology *Signs* 7, 4 (Summer 1982): 798-828.

Bibliography

PRIMARY

Unpublished Material and Archives

Association of MIT Alumnae. *Membership Survey–1972*. MIT Historical Collections, Institute Archives. Cambridge, MA.

Baltimore and Ohio Railroad Museum. Clipping File, Olive W. Dennis. Baltimore, MD.

Biographical Information File. Division of Engineering and Industry, National Museum of American History, Smithsonian Institution. Washington, D.C.

Bulletins of the Women's Bureau. *Records of the Women's Bureau, RG 86*. National Archives. Washington, D.C.

Bureau of Vocational Information Papers. Schlesinger Library, Radcliffe College. Cambridge, MA.

Collection of Materials on the MIT Women's Laboratory, 1875-1922. Institute Archives and Special Collections, MIT Libraries. Cambridge, MA.

Curtiss-Wright Corporation, 1942-1943. University Archives, University of Minnesota. Minneapolis, MN.

Curtiss-Wright Engineering Cadettes Program Papers. Archives of Women in Science and Engineering, Iowa State University. Ames, IA.

Dupont Papers. Hagley Museum and Library, Manuscripts and Archives. Wilmington, DE.

Ely Whitney Papers. Sterling Memorial Library, Manuscripts and Archives, Yale University. New Haven, CT.

Erasmus D. Leavitt. Diary, Erasmus D. Leavitt Jr. Collection, Division of Engineering and Industry, National Museum of American History, Smithsonian Institution, Washington D.C.

Frances Simpson Stevens Papers. Author's Personal Collection. Amsterdam, The Netherlands.

Frank and Lilian Gilbreth Papers. Department of Special Collections and Archives, Purdue University Library, Purdue University. Fayette, IL.

G.S. Morison Logbooks, 1863-1903. Dairies, Collected Papers of G. S. Morison, Division of Engineering and Industry, National Museum of American History, Smithsonian Institution, Washington, D.C.

Hall of History Foundation. Clipping File, General Electric Women. GE Schenectady, N.Y.

James Forgie Papers, 1890-1946. Division of Engineering and Industry, National Museum of American History, Smithsonian Institution. Washington, D.C.

Lou Hoover Papers. Herbert Hoover Presidential Library. West Branch, IA.
Mary Hallock Foote Papers. Special Collections, Stanford University. Stanford, CA.
Melville Kranzberg/SHOT Papers. Archives Center, National Museum of American History, Smithsonian Institution. Washington, D.C.
Mina Loy Papers. Beineke Manuscripts and Archives, Yale University. New Haven, CT.
Nora S. Blatch Papers. In Possession of Rhoda Barney Jenkins. Westport, CT.
Notebook Collection. Engineers' Notebook Collection, Housed Archives Center, National Museum of American History, Smithsonian Institution, Washington, D.C.
Ohio State University, Special Collections. Enrollment Figures. Ames, OH.
Photographic Collection. Division of Physical Sciences, National Museum of American History, Smithsonian Institution. Washington, D.C.
Ray Nathaniel Ray Correspondence, 1878-1916. New Haven Colony Historical Society. New Haven, CT.
The Society of Women Engineers Collection. Walter Reuther Library of Labor and Urban Affairs, Wayne State University. Detroit, MI.
Women Engineers. Clipping File, General Electric Co. Historical Collection, Hall of History. Schenectady, N.Y.
Women Engineers. Clipping File, by Ruth Bailey, General Electric Co.. Schenectady, N.Y.
Women Engineers. Clipping File, Schlesinger Library, Radcliffe College. Cambridge, MA.
Women Engineering Graduates in the U.S., 1877-1945 Papers. Oldenziel Archives, Amsterdam, the Netherlands

Interviews
Bott, Penny. "Steamboat Road and Nora Stanton Barney." Interview with John Barney and Rhoda Barney Jenkins. Greenwich: The Greenwich Library, 1978.
"Dorothy Quiggle." Author's Telephone Interview, November 14, 1989.
"Marion Monet." Author's Interview, March 23, 1990. Willow Street, PA.
"Melville Kranzberg." Author's Telephone Interview, August 31, 1990.
"Rhoda Barney Jenkins." Author's Interview, May 2, 1989. Greenwich, CT.

Published
Journals
291 (New York) (1915-6).
American Machinist ([1909-1920]).
American Society of Civil Engineers Proceedings.
California Journal of Technology (1903-4).
Cassier's Magazine.
Century.
Chemical and Metallurgical Engineering.
Engineering and Contracting.
Engineering News.
Harpers' Weekly.
Journal of the Society of Women Engineers (1951-4).
McClure's.
The Monad: A Magazine Devoted to the Social and Economic Welfare of the Engineer (1916-9).

The New Century for Women (1876). Newspaper at Centennial Exhibition.
One Big Union Monthly.
The Professional Engineer (formerly Monad) (1919-).
Scientific American.
Scribners.
SWE Newsletter (formerly Journal of the SWE) (1954-79).
Transactions of the American Institute of Mining and Mechanical Engineers.
The Woman Engineer (London) (1919-43).
Woman the Inventor (1890).

Books and articles

Abbott, Edith. *Women in Industry. A Study in American Economic History.* With an intro-duction by Sophonisba P. Breckinridge. New York and London: D. Appleton and Company, 1910.

Agle, William C. *In the Footsteps of Pizzarro: Or a Yankee's Five Years Experience in the Klondike of South America.* Seattle: Homer M. Hill, 1903.

Ainsworth, D[anforth] H[urlbut]. *Recollections of a Civil Engineer: Experiences in New York, Iowa, Nebraska, Dakota, Illinois, Missouri, Minnesota and Colorado.* Newton, Iowa: N.p., 1901 [1893].

"Air Engineer." *Christian Science Monitor,* 25 February 1943, 2.

"American Women Survey Their Emancipation: Careers Are Found an Aid to Successful Married Life." *Washington Post,* 16 August 1934.

Anderson, Benjamin M. "Technological Progress, The Stability of Business and the Inter-ests of Labor." Issued by the Chase National Bank of the City of New York. *The Chase Economic Bulletin* 17 (13 April 1937):3-35.

"Artistic Bridge and Highway Railings." *American City* 20 (March 1919):234-39.

Bailey, Ethel H. "Women as Engineers." *Independent Woman,* September 1932, 316-7, 334.

Baker, R. H., and Mary L. Reimold. "What Can Be Done to Train Women for Jobs in En-gineering." *Mechanical Engineering* 64 (December 1942):853 – 5.

"Barnard Girls Test Wireless 'phones'" *The New York Times,* 23 February 1909, 7: 3.

Barth, Emma C. "Engineers in Skirts." *National Business Woman,* February 1961, 6-7, 28.

Bates, Onward. *Bates, et al. of Virginia and Missouri. Onward Bates His Book.* Chicago: Press of P. F. Pettibone & Company, 1914.

–. "Onward and Onward: A Romance of Four Continents – memorabilia of an Engineer." Augusta, GA, 1933. mimeo by Ruth I. Hill.

Beach, Rex. *The Iron Trail: An Alaskan Romance.* New York: Harper and Brothers, 1913.

Beard, Mary Ritter. "Inventions Are for Men." Letter to the Editor. *The New York Times,* 17 August 1915, 8:8.

–. *Woman as Force in History.* New York: Macmillan, 1946.

–. *Woman's Work in Municipalities.* New York: Arno Press, 1972 [1915].

Bensel, John A. "Address at the 42d Annual Convention, Chicago, Illinois, June 21st, 1910." *Transactions of the American Society of Civil Engineers* 70 (December 1910):464-69.

Bigelow, Jacob. *Elements of Technology.* Boston, 1829.

Barney, Nora Stanton (Blatch). "Industrial Equality for Women." *N.Y. Herald-Tribune*, 21 April 1933.

–. "Wages and Sex." Letter to the Editor. *N.Y. Herald-Tribune*, 21 July 1933.

–. "Life Sketch of Elizabeth Cady Stanton." 1948.

–. *Women as Human Beings.* 1948.

–. *World Peace Through a People's Parliament, a Second House in World Government.* New York: Committee to Win World Peace Through a People's Parliament, 1944.

Boas, Franz. "The Occurrence of Similar Inventions in Areas Widely Apart." *Science* 9 (1887):485 – 6.

Bolanovich, D. J. "Selection of Female Engineering Trainees." *Journal of Educational Psychology,* December 1944, 545-53.

Bolanvich, J. "Selection of Female Engineering." *Journal of Educational Psychology,* December 1944, 545-53.

Bowden, Sue, and Avner Offer. "The Technological Revolution That Never Was: Gender, Class, and the Diffusion of Household Appliances in Interwar England." In *Sex of Things: Gender and Consumption in Historical Perspective,* Grazia Victoria de, and Ellen Furlough, 244-74. Berkeley: University of California Press, 1996.

Brady, Cyrus Townsend. *Web of Steel.* New York: Fleming H. Revell, 1916.

Brunton, David William. *Technical Reminiscences.* New York, 1915.

Bullard, F. F., arranged by. "A Son (?) of the M.I.T. [1897]." In *Technsongs: The M.I.T. Kommers Book,* Revised by committee of class of 1907 . Boston: Dibson, 1907 [1903].

Bureau of Vocational Information. *Women in Chemistry: A Study of Professional Opportunities.* New York: Bureau of Vocational Information, 1922.

Burgoyne, Arthur G. *Homestead: A Complete History of the Struggle Between the Carnegie Steel Company and the Amalgamated Association of Iron and Steel Workers.* Pittsburgh: Privately Printed, 1893.

Burr, Harriette. "Guidance for Girls in Mathematics." *The Mathematics Teacher* 36 (May 1943):203-11.

Burton, Richard F. "On the Lake Regions of Central Equatorial Africa." *Journal of the Royal Geographical Society (London)* 29 (1859).

Bush, Vannevar. *Science: The Endless Frontier.* A Report to the President. Washington, D.C.: U.S. Printing Office, 1945.

Carlyle, Thomas. "On History [1830]." In *The Varieties of History: From Voltaire to the Present,* Ed & comp & intro Fritz Stern. New York: Vintage, 1972 [1956].

Cartlidge, Oscar. *Fifty Years of Coalmining.* Oregon City: Oregon City Enterprise, 1933.

Carty, John J. "Ideals of the Engineer." Address Delivered at the Presentation of the John Fritz Meald at A.I.E.E. Convention, New York, NY, February 15, 1928. *Journal of the American Institute of Electrical Engineers* 47 (March 1928):210-12.

Cather, Willa. *Alexander's Bridge.* With an introduction by Sharon O'Brien. New York: Signet, 1987 [1912].

Century's Dictionary and Cyclopaedia. 1911.

Chaloner, Len. "Pioneering with Electricity." *Woman Citizen,* January 1926, 20.

Chappell, Eve. "Kate Gleason's Careers." *Woman Citizen,* January 1926, 19-20, 37-8.

Chase, Stuart. *Men and Machines.* New York, 1929.

–. *Men and Machines.* New York, 1929.

"Co-eds Engineers Take Men's Place." *Aviation News*, 27 December 1943, 13.

Collins, Francis Arnold. *The Fighting Engineers: The Minute Men of Our Industrial Army.* New York: Century, 1918.

Colvin, Fred H. *60 Years with Men and Machines: An Autobiography.* in collaboration with D. J. Duffin. New York: Whittlesey House, 1947.

Compton, Karl T. "Technology's Answer to Technocracy." In *For and Against Technocracy: A Symposium*, J. George ed. Frederick, 77-93. New York: Business Bourse, 1933.

Conroy, Stephen S. ":Thorstein Veblen's Prose." *American Quarterly* 20 (Fall 1968):605-15.

Cooley, Martimer E. *Scientific Blacksmith.* With Vivien B. Keatley. Ann Arbor: University of Michigan Press, 1947.

Cooney, Tom. *Meet Tom Cooney.* written in the first person by Mildred H. Comfort. Minneapolis: Lund Press, 1945.

Coyle, Juliet K. "Evolution of a Species – Engineering Aide." *U.S. Woman Engineer*, April 1984, 23-4.

Curwood, James Oliver. *The Danger Trail.* New York: Grosset and Dunlap, 1910.

Custer, Edgar A. *No Royal Road.* With a foreword by Samuel Vauclain. New York: K.C. Kinsey, 1937.

Daum, P. A. *H. Van Brakel, Ing. B.O.W.: oorspronkelijke roman door Maurits.* Leiden: A.W. Sijthoff, 1890.

Davis, Richard Harding. *Soldiers of Fortune.* Illustrated by C. D. Gibson. New York: Scribners, 1897.

Dennis, Olive W. "So – Your Daughter Wants to Be a Civil Engineer." *Baltimore and Ohio Magazine*, September 1940, 30.

Dixon, Merritt. *Sons of Martha: A Historical and Biographical Record Covering a Century of American Achievement by an Organization of Master Builders.* New York: Mason and Hanger, 1928.

Dorfman, Joseph. *Thorstein Veblen and His America.* New York: Vicking Press, 1934.

Dos Passos, John. *The Big Money.* U.S.A. Trilogy. New York: Washington Square Press, 1961 [1930].

Drews, Karl. "Women Engineers: The Obstacles in Their Way." *Scientific American. Supplement* 65 (7 March 1908):147 – 8.

Du Bois, W. E. B. *The Souls of Black Folk.* 1903.

Dubois, Ellen, ed. and introduced by. "Spanning Two Centuries: The Autobiography of Nora Stanton Barney." *History Workshop Journal* 22 (1986):131-52.

Eastman, Crystal. "Caroline Haslett and the Women Engineers." *Equal Rights* 11/12 (10 April 1926):69-70.

Eaves, Elsie. "Wanted: Women Engineers." *Independent Woman* 21 (May 1942):132-3, 158-9.

E[liot]., T. S. "Reviews, Kipling Redivivus." *Athenaeum*, 9 May 1919, 289.

Emmet, William LeRoy. *The Autobiography of an Engineer.* Albany: Fort Orange Press, 1931.

Etzler, John Adolphus. *The Paradise Within the Reach of All Men, Without Labor, by Powers of Nature and Machinery: An Address to All Intelligent Men.* Pittsburgh, 1833.

"Eyeball Engineer Draws 2d Glance." *Philadelphia Inquirer*, March 30 1970, 2.

Fackenthal Jr., Benjamin F[ranklin. "John Fritz the Iron Master." *The Pennsylvania German Society. Proceedings and Addresses* 34 (5 October 1929):97-112.

Folsom, Michael Brewster, and Steven D. Lubar, eds. *The Philosophy of Manufacturers: Early Debates Over Industrialization in the United States.* Documents in American Industrial History, Vol. 1. Cambridge, MA: MIT Press, 1982.

Foote, Elizabeth. "A Girl of the Engineers." *Atlantic Monthly* 95 (March 1905):381-91.

Foote, Mary Hallock. "In Exile." In *In Exile and Other Stories*, 1-58. Boston: Houghton, Mifflin, 1894.

–. *A Victorian Gentlewoman in the Far West: The Reminiscences of Mary Hallock Foote.* San Marino: Huntington Library, 1972.

Fox Jr., John. *The Trail of the Lonesome Pine.* New York: Scribner's Sons, 1908.

Fox, Francis. *63 Years of Engineering: Science and Social Work.* London: Murray, 1929.

Frances, L. "The Engineer." *Overland Monthly* 57 (January 1911):271-3.

Frederich, Augustine J., ed. *Sons of Martha: Civil Engineering Readings in Modern Literature.* New York: American Society of Civil Engineers, 1989.

Frederick, J. George, ed. *For and Against Technocracy: A Symposium.* New York: Business Bourse, 1933.

"Free-tuition in Courses Engineering for Women." *Science* 95 (9 January 1942):10.

Fritz, John. *The Autobiography of John Fritz.* New York: John Wiley & Sons, 1912.

"From a Sister of Martha." *Atlantic* 27 (March 1921):424-26.

The Gallup Pole: Public Opinion, 1935-1971. 3 Vols. New York: Random House, 1972.

G.S.B. "The Sons of Mary." *Literary Digest* 63 (8 November 1919):39.

Gage, Matilda E. Joslyn. *Woman as Inventor. Issued Under the Auspices of the New York State Woman Suffrage Association.* Woman Suffrage Tracts 1. New York, 1870.

–. "Woman as Inventor." *North American Review* 136 (May 1883):478-89.

Gebler, Robert T. "The Engineer (with Apologies to Kipling)." *Engineering News* 73 (14 January 1915):71.

Geddes, P. "Economic and Statistics, Viewed from the Standpoint of the Preliminary Sciences." Abstract of a Paper Presented at Section F of the British Association for the Advancement of Science, 1881. *Nature* 24 (29 September 1881):523-27.

Gibson, Louis H. "Art and Engineering." *Scientific American Supplement* 58 (October 1904):24033-4.

Gilbert, Alfred West. *Colonel A.W. Gilbert. Civil Engineer.* Cincinnati: Historical and Philosophical Society of Ohio, 1934.

–. *Bericht Über ein Allgemeines System von Abzugs-canälen Für die Stadt Cincinnati. Veröffentlicht Im Auftrage Des Stadtrathes.* Cincinnati: Buchdrückerei Des Demokratischen Tageblattes, 1852.

Gilbreth, Lillian M. "Women in Engineering." *Mechanical Engineering* 64 (1942):856-57, 859.

Gilfillan, S. C. "Social Effects of Inventions." In *Technological Trends and National Policy Including the Social Implications of New Inventions*, 24-66. National Resource Committee. Washington, D.C.: U.S. Government Printing Office, 1937.

–. *The Sociology of Invention.* Cambridge, MA: MIT Press, 1935.

"Girl Engineer in Kansas City." *Literary Digest* 78 (7 July 1923):29.

"Girls Studying Engineering See Future for Women in These Fields." *The Christian Science Monitor*, 19 April 1949, 10.

Goetzmann, William H. *Army Exploration in the American West, 1803-1863.* Yale Publications in American Studies, 4. New Haven: Yale University Press, 1959.

Goff, Alice C. *Women Can Be Engineers.* Youngstown, OH: Privately Printed, 1946.

–. "Women CAN Be Engineers." *AAUWJ* 41 (1948):75-76.

Grey, Zane. *The U.P. Trail.* New York: Harper and Brothers, 1918.

Groff, Frances A. "A Mistress of Mechanigraphics." *Sunset Magazine*, October 1911, 415 – 8.

H.M. "Kipling as Poet." *Poetry* 48 (April 1936):32-6.

Hancock, H. Irving. *The Young Engineers in Nevada.* Philadelphia: Henry Altemus, 1913.

Haney, James P. "Art and the Engineer: Combining the Useful with the Beautiful." *Scientific American Supplement* 71 (25 February 1911):119.

Hansen, Alvin. "Institutional Frictions and Technological Unemployment." *Quarterly Journal of Economics* 45 (August 1930-31):684-97.

Harland, Marion. "Martha and Her American Kitchen." *Woman's Home Companion*, March 1905, 12-3, 51, 55.

Harrison, Henry. "The Engineer and the Encyclopaedia." *Cassier's* 40 (December 1911):753-60.

Haslett, C. "Women's Contribution to Electrical Development." *Electrical Review* 99 (1926):655 – 6.

Hazen, Edward. *Popular Technology: Or, Professions and Trades.* New York: Harper and Brothers, 1841.

Herringshaw, Thomas W. *Herringshaw's National Library of American Biography.* Vols. 5. Chicago: American Publishers Association, 1909-14.

Hine, Lewis W. *Men at Work.* New York: Dover, 1977 [1932]

Hitchcock, Embury A. *My Fifty Years in Engineering.* Caldwell, Idaho: Caxton Printers, 1939.

Holcomb, Walter. *Memories of Walter Holcomb of Torrington, Litchfield County, Connecticut with a Few Departures in Genealogy, Public Records, Customs Etc.* N.p. 1935.

Horneman, Beatrice. "Engineers Wanted." *Independent Woman*, July 1950.

Howard, Robert Milton. *Reminiscences.* Columbus, GA.: Gilbert Press, 1912.

Howe, E. D. "Training Women for Engineering Tasks." *Mechanical Engineering* 65 (October 1943):742 – 3.

Huet, Conrad Busken. *Lidewyde.* Den Haag: Marinus Nijhoff, 1981 [1868].

Huygens, Cornélie. *Barthold Meryan.* Amsterdam: P. N. Van Kampen & Zoon, n.d.

"The Industrial Encyclopedia." *The One Big Union Monthly* 1 (December 1919):15.

Ingels, Margaret. "Petticoats and Slide Rules." *Midwest Engineer* (1952).

"Jersey Girl Has Job as Shipyard Engineer." *The New York Times*, 8 May 1949, 29.

Johnson, James. "Women Inventors and Discovers." *Cassier's Magazine. An Engineering Magazine* 36 (October 1909):548-52.

Kersey, John. *Dictionarium Anglo-Britannicum.* London: Phillips, 1708.

Kimball, Dexter S. *I Remember.* New York: McGraw-Hill, 1953.

Kipling Rudyard. *Just So Stories.* London: Macmillan, 1902.

–. "The Bridge Builders." In *A Day's Work.* New York: Doubleday and McCure, 1898.

–. *Captains Courageous: A Story of the Grand Banks.* New York: Macmillan, 1897.

–. *Many Inventions.* London: Macmillan, 1893.

–. "The Sons of Martha." *The Standard (London),* 29 April 1907.

–. "Sons of Martha." with an introduction by Arthur M. Lewis. Chicago: John F. Higgins, n.d.

–. "The Sons of Martha." Illustrated by Winter. *The New York Tribune,* 28 April 1907.

–. "The Sons of Martha." *Philadelphia Press,* 28 April 1907.

–. *The Sons of Martha.* Broadside. Garden City: Doubleday, Page, 1907.

"Lady Auto Engineer: Her Ideas Irreparable." *New York World-Telegram & Sun,* 14 August 1961.

Lamme, Benjamin Garver. *Electrical Engineer: An Autobiography.* New York: Putnam, 1926.

Lane, Ann J. *Mary Ritter Beard: A Sourcebook.* New York: Schocken Books, 1977.

Lawton, R[alph] W[aldo]. *An Engineer in the Orient.* Los Angeles: Walton & Wright, 1942.

Leffingwell, Georgia. "A Weather Engineer." *Woman's Journal* (1930):22-3.

Leonard, John W., ed. *Who's Who in Engineering: A Biographical Dictionary of Contemporaries 1922-1923.* New York: John W. Leonard, 1922.

–. *Who's Who in New York City and State Containing Authentic Biographies of New Yorkers.* New York: L.R. Hamersly, 1907.

Leuck, Miriam Simons. "Women in Odd and Unusual Fields of Work." *AAAPPS* 143 (1929):166-79.

Lindsay, Malvina. "The Gentler Sex: Young Women in a Hurry." *Washington Post,* 20 July 1944.

Litchfield, Paul Weeks. *Autumn Leaves: Reflections of an Industrial Lieutenant.* Cleveland: The Corday & Gross, 1945.

Little, Arthur D. "The Fifth Estate." Address Delivered at Franklin Institute Centenary. *Atlantic Monthly* 134 (December 1924):771-81.

–. "Technocracy vs. Technology." *Commercial and Financial Chronicle* 136 (21 January 1933):435 – 7.

Littlepage, John D. *In Search of Soviet Gold.* In collaboration with Demaree Bess. New York: Harcourt, Brace and Company, 1938.

Lydgate, William A. "The 32 Million-word Classic." *Saturday Review* 25 (2 May 1942 1942):3-4, 22-3.

Lynde, Francis. *The Quickening.* Indianapolis: Bobbs-Merrill, 1906.

McCarty, Richard Justin. *Work and Play.* Kansas City: Empire Press, 1912.

Manson, L. "America Gave Me Opportunity." *Power Plant Engineering* 19 (April 1945):64, 130.

"Margaret Ingels of Carrier Honored as 'pioneer' in Women's Career." *Refrigerating Engineering* 41 (January 1941):62.

Mason, Otis Tufton. "The Birth of Invention." In *Patent Centennial Celebration. Proceedings and Address. Celebration of the Beginning of the Second Century of the American Patent System at Washington City, DC, April 8, 9, 10, 1891,* 403-12. Washington, D.C.: Press of Gedney & Roberts, 1892.

–. "The Occurrence of Similar Inventions in Areas Widely Apart." *Science* 9 (1887):534-35.

–. "Woman as an Inventor and Manufacturer." *Popular Science Monthly* XLVII (May-October 1895):92-103.

–. *Woman's Share in Primitive Culture.* Anthropological Series, 1. New York: D. Appleton, 1894.

Massachusetts Institute of Technology. *Reports of the President, Secretary and Departments, 1871-1872.* Boston: Press of A. A. Kingman, 1872.

Matschat, Cecile Hulse. *Seven Grass Huts: An Engineer's Wife in Central-and-South America.* New York: Farrar & Rinehart, 1939.

Mencken, Henry L. "Professor Veblen and the Cow." *Smart Set* 59 (May 1919):138-44.

Merrill, Dot. "But You Don't Look Like an Engineer." in collaboration with Richard B. Espey. *National Business Woman,* May 1953, 4-5, 15.

Merritt, Dixon. *Sons of Martha: A Historical and Biographical Record Covering a Century of American Achievement by an Organization of Master Builders.* New York: Mason and Hanger, 1928.

Merritt, Raymond H. *Engineering in American Society, 1850-1875.* Lexington: The University Press of Kentucky, 1969.

"Miss Clare Nicolet is Right at Home Among Turbines in Powerhouse of Kansas City Railways Company." *Aera* 11 (1923):1456-8.

Moore, Colleen. *Silent Star.* New York: Doubleday, 1968.

Moorhouse, C. E. "Mr Wardrop's Problem: Excerpts from a Talk on Kipling and Technology." *Kipling Journal* 61 (March 1987):10-22.

Morgan, Lewis Henry. *Ancient Society: Or, Researches in the Lines of Human Progress from Savagery Through Barbarism to Civilization.* With a foreword by Elizabeth Tooker. Classics of anthropology. Tuscon: University of Arizona, 1985 [1877]

Morrow, Honoré (McCue) Willsie. *The Enchanted Canyon.* New York: A. L. Burt, 1921.

–. *The Heart of the Desert.* New York: A.L. Burt, 1913.

–. *Still Jim.* New York: Frederick A. Stokes, 1914.

"Mrs. De Forest Files Suit." *The New York Times,* 1 January 1916, 18.

"Mrs. De Forest Loses Suit." *The New York Times,* 22 January 1916, 13.

"Mrs. De Forest, Suing, Tells Her Real Age." *The New York Sun,* 1 January 1916, 7.

Mumford, Lewis. "If Engineers Were Kings." *The Freeman* 4 (23 November 1921):261-62.

New Americanized Encyclopaedia Britannica. [1896-1904].

New Encyclopaedia Britannica. Chicago: Encyclopaedia Britannica, [1974-1968].

Objects and Plan of an Institute of Technology; Including a Society of Arts, a Museum, and a School of Industrial Science. Boston: John Wilson and Son, 1861 [second ed.].

Ogburn, William F. *Living with Machines.* Chicago: American Library Association, 1933.

–. "National Policy and Technology." In *Technology and Society. The Influence of Machines in the U.S.* 1937.

–. "National Policy and Technology." In *Technological Trends and National Policy Including the Social Implications of New Inventions,* National Resources Committee, 3-14. Washington, D.C.: U.S. Government Printing Office, 1937.

–. with S.C. Gilfillan "The Influence of Invention and Discovery." In *Recent Social Trends in the United States.* New York: McGraw-Hill, 1933.

"Old Men Bar Miss Blatch." *The New York Times,* 12 January 1916, 7.

Parsons, Rachel M. "Engineering as a Profession for Women." *Living Age* 304 (10 January 1920):116-21.

Patent Centennial Celebration. *Proceedings and Addresses. Celebration of the Beginning of Teh Seocnd Century of the American Patent System at Washington City, DC, April 8, 9, 10, 1891.* Published by Executive Committee. Washington, D.C.: Press of Gedney & Roberts, 1892.

Pedrick, Howard Ashley. *Jungle Gold: Dad Pedrick's Story.* Collaborators Will de Grouchy and L. Magee. New York: Bobb Merrill, 1931.

"Personalities in Industry." *Scientific American* 264 (April 1941):197.

Pickel, Margaret Barnard. "A Warning to the Career Women." *The New York Times Magazine*, 16 July 1944, 19, 32-3.

Pickering, John. *A Vocabulary or Collection of Words and Phrases Which Have Been Supposed to Be Peculiar to the United States Etc.* Boston: Cummings and Hilliard, 1816.

Pickett, William D. *A Sketch of the Professional Life of William D. Pickett of Four Bear, Wyoming.* Louisville: John P. Morton, 1904.

Poole, Ernest. *The Harbor.* New York: Macmillan, 1915.

Porter, Charles T. *Engineering Reminiscences.* New York: Wiley and Sons, 1912.

Pound, Arthur. *The Turning Wheel: The Story of General Motors Through Twenty-Five Years, 1908-1933.* Garden City, NY: Doubleday, Doran & Company, 1934.

Proceedings of the Encyclopedia Britannica Conference on the Technological Order. Edited by Carl F. Stover. Detroit: Wayne State University Press for the Society of the History of Technology, 1962.

Pupin, Michael Idvorsky. *From Immigrant to Inventor.* New York: Scribner's, 1923.

Randolph, Isham. *Gleanings from a Harvest of Memories.* Columbia: E.W. Stephens, 1937.

Ray, Anna Chapin. *The Bridge Builders.* Boston: Little, Brown, 1909.

Renshaw, Clarence, lyricist. LLoyd B. Haworth, composer. "Technology." In *Techsongs: The M.I.T. Kommers Book*, revised by . Boston: O. Dibson, 1907 [1903].

Reuleaux, Franz. "Technology and Civilization." In *Annual Report of the Board of Regents of the Smithsonian Institution Showing the Operations, Expenditures, and Condition of the Institution to July, 1890. General Appendix.*, 705-17. Washington, DC: Government Printing Office, 1891.

Reynolds, Minnie J. "Women as Inventors." Interurban Woman Suffrage Series no 6. New York: Interurban Woman Suffrage Council [Reprint New York Sun, October 25, 1908], 1908.

Ribinskas, Jeanne M. "The Distaff Engineer." *Automation*, May 1974, 50-55.

Richards, Robert Hallowel. *His Mark.* Boston: Little, Brown, 1936.

Rickard, Thomas Arthur. *A Guide to Technical Writing.* San Francisco: Mining and Scientific Press, 1908.

–. *Technical Writing.* New York: John Wiley & Sons, 1920.

Ridgway, Robert. "My Days of Apprenticeship." *Civil Engineering* 8 (September 1938):601-4.

–. *Robert Ridgway.* with Isabelle Law Ridgway. New York: Privately Printed, 1940.

Rolt, L. T. C. *A Short History of Machine Tools.* Cambridge, MA: MIT Press, 1965.

Roosevelt, Theodore. "The Strenuous Life: Speech Before the Hamilton Club." In *The Strenuous Life: Essays and Addresses.* New York: Century, 1902 [1899].

Root, Henry. *Henry Root, Surveyor, Engineer, and Inventor.* San Francisco: Privately Printed, 1921.

Rugg, Harold Ordway. *The Great Technology: Social Chaos and the Public Mind.* New York: The John Day Company, 1933.

Sauveur, Albert. *Metallurgical Reminiscences.* New York: American Institute of Mining and Metallurgical Engineers, 1937.

Scott, Howard. *Introduction to Technocracy.* New York: The John Day Company, 1933.

–. "Technocracy Speaks." *Living Age* 343 (December 1932):297-303.

Scott, R. G. "Memoirs and Poetic Sketches." N.p. Reveille Print.

Searles, James M. *Life and Times of a Civil Engineer; Supplemented by the True Theory of the Mississippi River.* Cincinnati: Robert Clarke, 1893.

Seligman, Edwin R. A., and Alvin Johnson, eds. *Encyclopaedia of the Social Sciences.* New York: Macmillan, 1934.

"She Debates Plane Design Just Like Recipe for Pie." *Toronto Daily Star,* 8 February 1940.

Shelmire, Robert W. *The Draftsman.* Chicago: Chicago Publishing Bureau, 1919.

Shute, Nevil. Helmut Ackermann, illustrator. *Slide Rule: The Autobiography of an Engineer.* London: Heron Books, 1968.

Sloan, Alfred P. *My Years with General Motors.* Edited by John McDonald. With an introduction by Catharina Stevens and Peter F. Drucker. New York: Doubleday, 1990 [1963].

Smallwood, Julian Chase. "Engineering and Art." *Cassier's* 37 (January 1910):213-6.

Smiles, Samuel. *Selections from Lives of the Engineers, with an Account of Their Principal Works.* Ed and introd by Thomas Parke Hughes. Cambridge, MA: MIT Press, 1966.

–. *Self-Help.* London. 1859.

Spearman, Frank H. *Whispering Smith.* New York: Charles Scribner's Sons, 1906.

Stanton, Elizabeth Cady, Susan B. Anthony, and Matilda Joslyn Gage. *History of Woman Suffrage.* Vol. 3 (1876-1885). Rochester: Charles Mann, 1886.

Starret, Paul. *Changing the Skyline.* New York: Wittlesley House, 1938.

Steele, Evelyn M. *Career for Girls in Science and Engineering.* New York: E.P. Dutton, 1943.

Stegner, Wallace. *Angle of Repose.* New York: Doubleday, 1971.

Steinman, David B. "Engineers Pay Tribute to the Woman Who Helped Build the Brooklyn Bridge." Address for the Brooklyn Engineers Club, May 24, 1953. *The Transit of Chi Epsilon,* Spring-Fall 1954, 1-7.

Stevens, John Frank. *An Engineer's Recollections.* New York: McGraw-Hill, 1936.

Stocking, George W., ed. *A Franz Boas Reader: The Shaping of American Anthropology, 1883-1911.* Chicago: The University of Chicago Press, 1974.

"Structural Art." *Literary Digest* 67 (25 December 1920):26-7.

Stuart, Charles B. *Lives and Works of Civil and Military Engineers of America.* New York: Van Nostrand, 1871.

Talmadge, F. M. "Engineering Training for Women." *Journal of Higher Education,* October 1940, 379-82.

Tarbell, Ida M. "Women as Inventors." *Chautauquan* 7 (March 1887):355-57.

Taylor, Frederick Winslow. *The Principles of Scientific Management.* New York: W. W. Norton, 1967 [1911]

"Technology of Ceramic Art." *The New Century for Woman* 23 (24 Saturday October 1876):179.

Thoreau, Henry David. "Paradise (to Be) Regained." *United States Magazine and Democratic Review* 13 (November 1843):451-63.

"Training for Women in Aeronautical Engineering at the University of Cincinnati." *Science* 97 (18 June 1943):548 – 9.

Training in Business and Technical Careers for Women. Ohio: The University of Cincinnati, 1944.

Traylor, Samuel W. *Out of the South West: A Texas Boy.* Allentown, PA: P. Schlicher & Son, 1936.

Tribune. *Tribune Guide to the Exhibition.* New York: The Tribune, 1876.

Tuft, Edward M. "Women in Electronics." *National Business Women*, November 1956.

U.S. Bureau of Labor Statistics. "Effects of Technological Changes Upon Employment in the Motion-Picture Theaters of Washington, DC." *Monthly Labor Review* 33 (November 1931):1005-18.

U.S. Bureau of the Census. *Abstract to the Eleventh Census.* Washington, D.C. 1890.

–. "*Census.*" Washington, D.C.: Government Printing Office, [1890-1940].

–. *Special Reports. Occupations at the Twelfth Census.* Washington, D.C.: Government Priting Office, 1904.

–. *Statistics of Women at Work.* Based on Unpublished Information Derived from the Schedules of the Twelfth Census 1900. Washington, D.C.: Government Printing Press, 1907.

–. *Twenty Censuses: Population and Housing Questions, 1790-1980.* Washington, D.C.: Government Printing Office, 1979.

U.S. Department of Interior, Office of Education. *Land-Grant Colleges and Universities.* Washington: Government Printing Office, 1930.

U.S. Department of Labor, Woman's Bureau. *The Occupational Progress of Women, 1910 to 1930.* Bulletin No. 104. Washington, D.C.: Government Printing Office, 1933.

–. *The Outlook for Women in Architecture and Engineering.* Bulletin 223 no.5. Washington, D.C.: Government Printing Office, 1948.

–. *Women's Contribution in the Field of Invention.* Bulletin Vol. 28. Washington, D.C.: Government Printing Office, 1923.

–. "Employment and Characteristics of Women Engineers." *Monthly Labor Review*, May 1956, 551-6.

U.S. Office of Education, Department of Interior. "Survey of Land-Grant Colleges and Universities." Washington, D.C.: U.S. Government Printing Office, 1930.

U.S. Patent Office. *Women Inventors to Whom Patents Have Been Granted by the United States, 1790 to July 1, 1888.* With Appendices to March 1, 1895. Washington, D.C.: Government Printing Office, 1895.

U.S. Bureau of Labor Statistics. "Effect of Technological Changes Upon Employment in the Amusement Industry." *Monthly Labor Review* 32 (August 1931):261-7.

United States Centennial Commission. International Exhibition, 1876. *Reports and Awards.* Edited by Francis A. Walker. Washington, D.C.: Government Printing Office, 1880.

United States Directory of Registered Professional Engineers: A Biennial Publication. New York: Lewis Historical Publishing Company, 1937-.

Usher, Abbott Payson. *A History of Mechanical Inventions.* New York: McGraw-Hill, 1929.

Vaile, Charlotte M. *The M.M.C.: A Story of the Great Rockies.* Illustrated by Sears Gallagher. Boston: W. A. Wilde, 1898.

Van Deusen, E. A. "The Engineer in Fiction." *The Professional Engineer,* September 1922, 15.

Veblen, Thorstein. *The Engineers and the Pricesystem.* New York: B. W. Huebsch, 1921.

–. *The Instinct of Workmanship.* New York, 1914.

–. "The Place of Science in Modern Civilization." *The American Journal of Sociology* 11 (March 1906).

–. *The Theory of the Leisure Class: An Economic Study of Institutions.* With an introduction by Stuart Chase. New York: Modern Library, 1934 [1899]

–. "An Unpublished Paper on the I.W.W. by Thorstein Veblen." with an introduction by Joseph Dorfman. *Journal of Political Economy* 40 (December 1943):796-807.

–. "Using the I.W.W. to Harvest Grain." Unpublished Paper (1918). introd by Joseph Dorfman. *Journal of Political Economy* 40 (December 1932):796-807.

Venable, Matthew Walton. *Eighty Years After: Or Grandpa's Story.* Charleston: Hood-Heiserman-Brodhag, 1929.

Waddell, J. A. L. *Memoirs and Addresses of Two Decades by J. A. L. Waddell, Consulting Engineer.* Edited by Frank W. Skinner. Easton, PA: Mack Printing, 1928.

Walker, Francis A. "The Place of Schools of Technology in American Education." *Educational Review* (1891):209-19.

Walker, Timothy. "Defense of Mechanical Philosophy." *North American Review* 31 (July 1831):122-36.

Warner, Deborah J. "Women Inventors at the Centennial." In *Dynamos and Virgins Revisited,* Martha Moore Trescott, 102-19. Metuchen, NJ: Scarecrow Press, 1979.

"Warns Wives of 'careers'" *The New York Times,* 31 July 1911, 6:5.

Washington, Booker T. *Up from Slavery.* New York: A.L. Burt, 1901.

–. "Industrial Education." *Annals of American Academy of Political and Social Science* 33, no. 1 (1909).

Webster, Noah. *American Dictionary of the English Language.* New York: Converse, 1828.

Whitnach, Donald R., ed. *Government Agencies.* Greenwood Encyclopedia of American Institutions. Westport, Ct.: Greenwood Press, 1983.

Who's Who in Engineering: A Biographical Dictionary of the Engineering Profession. Edited by J. W. Leonard. New York: Lewis Historical Publishing, 1922-64.

Who's Who in Engineering. New York: Lewis Historical Publishing, 1941 [5th edition].

Wickenden, William E. *Report of the Investigation of Engineering Education.* Pittsburgh: Society for the Promotion of Engineering Education, 1930.

Wilson Cole, W. "Training of Women in Engineering." *Journal of Engineering Education* 43 (October 1943):167-84.

Wilson, William Hasell. *Reminiscences of William Hasell Wilson.* Edited by Elizabeth B. Pharo. Philadelphia: Patterson and White, 1937.

"Woman in Engineering Class at University of Maryland." *Sun,* Saturday, 27 September 1941.

"Woman's Inferiority to Man in Light of Engineering Science." *Current Literature* 44 (May 1908):553 – 4.

"Women AIME Members Contribute Their Share in Engineering War." *Mining and Metallurgy* 23 (November 1942):580 – 1.

"Women Engineers-yesterday and Today." *The Bent of Tau Beta Pi*, Summer 1971, 10-2.

"Women Sought in Engineering." *AAUWJ* 36 (1943):93-94.

Worrall, James. *Memoirs of Colonel James Worrall, Civil Engineer, with an Obituary Postscript by a Friend.* Harrisburg, PA: E.K. Meyer Printed, 1887.

—. *Report of a Survey of South Pennsylvania Railroad.* n.p. 1862.

—. *Report of the Pennsylvania Board of Pennsylvania and New York Joint Boundary Commission.* Harrisburg, PA: L.S. Hart, State Printer, 1878.

Wright, Harold Bell. *The Winning of Barbara Worth.* Chicago: Book Supply, 1912.

Wright, Willard Huntington [pseud. for S. S. Van Pine]. *Misinforming the Nation.* New York: B. W. Huebsch, 1917.

Yost, Edna. *American Women of Science.* New York: Frederick A. Stokes, 1943.

—. *Frank and Lillian Gilbreth: Partners for Life.* New Brunswick: Rutgers University Press, 1949.

SECONDARY

Published
Books and articles

100 Years: A Story of the First Century of Vanderbilt University School of Engineering 1875-1975. Nashville: Vanderbilt Engineering Alumni Association, 1975.

A.A.C. "In Memoriam: Anna Chapin Ray." *Smith Alumnae Quarterly* 37 (February 1946):87.

Abbott, Andrew. *The Systems of Profession: An Essay on the Division of Expert Labor.* Chicago: Chicago University Press, 1988.

Akin, William E. *Technocracy and the American Dream: The Technocracy Movement, 1900-1941.* Berkeley: University of California Press, 1977.

American Association of Engineering Societies. "Women in Engineering." *Engineering Manpower Bulletin* 99 (December 1989).

Ammons, Elizabeth. "The Engineer as Cultural Hero and Willa Cather's First Novel, Alexander's Bridge." *American Quarterly* 38 (Winter 1986):746-60.

"Art Into Life: Russian Constructivism 1914-1932." Exhibit Walker Art Center, Minneapolis, October 7-December 30, 1990. 1990.

Bailyn, Lotte. "Experiencing Technical ork: A Comparison of Male and Female Engineers." *Human Relations* 40, no. 5 (1987):299-312.

Baker, Paula. "The Domestication of Politics: Women and American Political Society, 1780-1920." *American Historical Review* 89 (June 1984):620-47.

—. *The Moral Framework of Public Life: Gender, Politics and the State in Rural New York, 1870-1920.* New York: Oxford University Press, 1991.

Baron, Ava. "Contested Terrain Revisited: Technology and Gender Definitions of Work in the Printing Industry, 1850-1920." In *Women, Work, and Technology. Transformations,* edited by Barbara Drygulski Wright, 58-83. Ann Arbor: The University of Michigan Press, 1987.

–. "An 'Other' Side of Gender Antagonism at Work: Men, Boys, and the Remasculinization of Printer;s Work, 1830-1920." In *Work Engendered: Towards a New History of American Labor*, 1-46. Ithaca, NY: Cornell University Press, 1991.

Baumer, Franklin L. *Modern European Thought: Continuity and Change in Ideas, 1600-1950*. New York: Macmillan, 1977.

Bederman, Gail. *Manliness & Civilization: A Cultural History of Gender and Race in the United States, 1880-1917*. Chicago: The University of Chicago Press, 1995.

Bell, Daniel. "Veblen and the Technocrats: On The Engineers and the Price System." In *The Winding Passage: Essays and Sociological Journeys, 1960-1980*, 69-90. New York: Basic Books, 1980 [1963].

Belofsky, Harold. "Engineering Drawing – a Universal Language in Two Dialects." *Technology and Culture* 32 (January 1991):23-46.

Bennett, Stuart. "'Industrial Instrument-master of Industry, Servant of Management': Automatic Control in the Process Industries, 1900-1940." *Technology and Culture* 32 (January 1990).

Berteaux, Daniel, ed. *Biography and Society: The Life History Approach in the Social Sciences*. Sage Studies in International Sociology. Beverly Hills: Sage, 1981.

Bieder, Robert E. *Science Encounters the Indian, 1820-1880: The Early Years of American Ethnology*. Norman: The University of Oklahoma Press, 1986.

Black, Max. "More About Metaphor." In *Metaphor and Thought*, edited by Andrew Ortony, 19-43. Cambridge: Cambridge University Press, 1979.

Blank, David M., and George J. Stigler. *The Demand and Supply of Scientific Personnel*. National Bureau of Economic Research, General Series, 62. New York: National Bureau of Economic Research, 1957.

Bledstein, Burton. *The Culture of Professionalism*. New York: Norton, 1976.

Blumin, Stuart M. *The Emergence of the Middle Class: Social Experience in the American City, 1760-1900*. Cambridge: Cambridge University Press, 1989.

–. "The Hypothesis of Middle-class Formation in Nineteenth-century America." *American Historical Review*, April 1985, 299-328.

Boydston, Jeanne. *Home and Work: Housework, Wages and the Ideology of Labor in the Early Republic*. New York: Oxford University Press, 1990.

–. "'To Earn Her Daily Bread': Housework and Antebellum Working-class Subsistence." *Radical History Review* 35 (April 1986):7-25.

Branfon, Robert L. *Cotton Kingdom of the New South: A History of the Yazoo – Mississippi Delta from Reconstruction to the Twentieth Century*. Cambridge, MA: Harvard University Press, 1967.

Brecher, Jeremy. *Strike!* Boston: South End Press, 1972.

Brod, Harry, ed. *The Making of Masculinities*. Boston: Allen and Urwin, 1987.

Brody, David. *Steelworkers in America: The Nonunion Era*. New York: Harper Torchbooks, [1960].

Brown, John K. *The Baldwin Locomotive Works, 1831-1915: A Study of American Industrial Practice*. Baltimore: The Johns Hopkins University Press, 1995.

Brumberg, Joan Jacobs, and Nancy Tomes. "Women in the Professions: A Research Agenda for American Historians." *Reviews of American History* 10 (June 1982):275-96.

Bugliarello, George, et al (eds.). *Women in Engineering: Bridging the Gap Between Society and Technology.* Proceedings of an Engineering Foundation Conference. Chicago: University of Illinois, 1971.

Burke, Carolyn. *Becoming Modern. The Life of Mina Loy.* Berkeley: University of California Press, 1996.

Burke, Carolyn, and Naomi Sawelson-Gorse. "In Search of Frances Simpson Stevens." *Art in America*, April 1994.

Burkett, Eva Mae. *American Dictionaries of the English Language Before 1861.* Metuchen, NJ: Scarecrow Press, 1979.

Burks Esther Lee. "Career Interruptions and Perceived Discrimination Among Women in Engineering and Science." In *Women in Engineering*, Mary Ott, and Nancy A. eds. 1975.

Calhoun, Daniel Hovey. *The American Civil Engineer: Origin and Conflict.* Cambridge, MA: MIT Press, 1960.

Canel, Annie. "Following the Trail of the Pioneers: Women Engineers in the French 'Grandes Ecoles' in the 1960s and 1970s." *History and Technology* 14 (January 1997):123-46.

Carlyle, Thomas. "On History [1830]." In *The Varieties of History: From Voltaire to the Present*, Ed & comp & intro Fritz Stern. New York: Vintage, 1972 [1956].

Carnes, Mark C. *Secret Ritual and Manhood in Victorian America.* New Haven: Yale University Press, 1989.

Carnes, Mark C., and Clyde Griffen, eds. *Meanings for Manhood: Constructions of Masculinity in Victorian America.* Chicago: The University of Chicago Press, 1990.

Cawelti, John G. *Apostles of the Self-made Man.* Chicago: University of Chicago Press, 1965.

Chandler Jr., Alfred D. *The Visible Hand: The Managerial Revolution in American Business.* Cambridge: Harvard University Press, 1977.

Chaplin, Ralph. *Wobbly: The Rough-and-tumble Story of an American Radical.* New York: Da Capo Press, 1972 [1948].

Chase, Stuart. "Waste and Labor." *Nation* 112 (20 July:67-69.

Christensen, Lawrence O. "Being Special: Women Students at the Missouri School of Mines and Metallurgy." *Missouri Historical Review* 83 (October 1988):17-35.

Christensen, Lawrence O., and Jack B. Ridley. *UM-Rolla: A History of MSM/UMR.* University of Missouri Printing Services, n.d.

Clark, Jennifer. "The American Image of Technology from the Revolution to 1840." *American Quarterly* 39 (Fall 1987):431-49.

Cockburn, Cynthia. *Machinery of Dominance. Women, Men, and Technical Know-how.* foreword by Ruth Schwartz Cowan. Boston: Northeastern University Press, 1988 [1985]

Cockburn, Cynthia, and Susan Ormrod. *Gender & Technology in the Making.* London: Sage, 1993.

Colburn, Warren. "Teaching of Arithmetic." Boston Address Before the American Institute of Instruction, 1830. In *Readings in the History of Mathematics Education*, 24-37. Washington, D.C.: National Council of Teachers in Mathematics, 1970.

Cott, Nancy F., ed. and introduction by. *A Woman Making History: Mary Ritter Beard Through Her Letters.* New Haven: Yale University Press, 1991.

−. *The Bonds of Womanhood: "woman's Sphere" in New England, 1780-1835.* New Haven: Yale University Press, 1977.

−. *The Grounding of Modern Feminism.* New Haven: Yale University Press, 1987.

Cragg, Barbara. "Mary Hallock Foote's Images of the Old West." *Landscape* 24, no. 3 (1980):42-7.

Cutcliffe, Stephen H., and Robert C. Post, eds. *In Context: History and the History of Technology. Essays in Honor of Melville Kranzberg.* Bethlehem: Lehigh University Press, 1989.

d'Harnoncourt, Anne. *Futurism and the International Avant-garde.* Catalogue to Exhibition, October 26, 1980 to January 4, 1981. Philadelphia: Philadelphia Museum of Art, 1980.

Dabakis, Melissa. "Douglas Tilden's *Mechanics Fountain*: Labor and the 'Crisis of Masculinity' in the 1890s." *American Quarterly* 47 (June 1995):204-35.

Davidson, Cathy N., ed. "Reading in America. Literature & Social History." Baltimore: The Johns Hopkins University Press, 1989.

−. *Revolution and the Word: The Rise of the Novel in America.* New York: Oxford University Press, 1986.

Davis, David B. *From Homicide to Slavery: Studies in American Culture.* New York: Oxford University Press, 1987.

Davis, Nathalie Zemon, and Randolph Starn, eds. "Memory and Counter-Memory. Special Issue." *Representations* 26 (1989):1-6.

Derber, Charles, ed. *Professionals as Workers: Mental Labor in Advanced Capitalism.* Boston: G.K. Hall, 1982.

Diggins, John P. *The Bard of Savagery. Thorstein Veblen and Modern Social Theory.* New York: The Seabury Press, 1978.

Douglas, Ann. *The Feminization of American Culture.* New York: Alfred A. Knopf, 1977.

Douglas, Susan J. *Inventing American Broadcasting, 1899-1922.* Baltimore: The Johns Hopkins University Press, 1987.

Dresselhaus, M. S. "Reflections on Women Graduate Students in Engineering." *IEEE Transactions on Education* 28 (November 1985):196-203.

Edge, D. O. "Technological Metaphor." In *Meaning and Control. Essays in Social Aspects of Science and Technology,* edited by D. O. Edge, edited by J. N. Wolfe, 31-59. London: Tavistock Publications Limited, 1973.

Eichhorn, Robert L. "The Student Engineer." In *The Engineers and the Social System,* eds. Robert Perrucci and Joel Gerstl. New York: John Wiley and Sons, 1969.

Fee, Elizabeth. "The Sexual Politics of Victorian Social Anthropology." In *Clio's Consciousness Raised,* Eds Mary S. Hartman and Lois W. Banner, 86-102. New York, 1976.

Feibleman, James. "Pure Science, Applied Science, Technology, Engineering: An Attempt at Definitions." *Technology and Culture,* Fall 1961, 305-17.

Ferguson, Eugene S. "Expositions of Technology, 1851-1900." In *Technology in Western Civilization,* Eds. Melvin Kranzberg and Carroll Jr. Pursell, 706-26. New York: Oxford University Press, 1967.

−. "Kinematics of Mechanisms from the Time of Watt." *Contributions from the Museum of History and Technology* (1963):186-230.

−. "The Mind's Eye: Nonverbal Thought in Technology." *Science,* 26 August 1977, 827-36.

−. "Power and Influence: The Corliss Engine in the Centennial Era." In *Bridge to the Future: A Centennial Celebration of the Brooklyn Bridge*, Margaret Latimer, et al., 225-46. Annals of the New York Academy of Sciences 424. New York: New York Academy of Sciences, 1984.

−. "Technical Journals and the History of Technology." In *In Context: History and the History of Technology. Essays in Honor of Melville Kranzberg*, Eds Stephen H. Cutcliffe and Robert C. Post, 312-13. Bethlehem: Lehigh University Press, 1989.

Fijalkowski, Krzysztof. "Dada and the Machine." *Journal of European Studies* 17 (1987):233-51.

Fink, Gary M. *Biographical Dictionary of American Labor*. Westport, CT: Greenwood Press, 1984.

−. ed. *Labor Unions*. The Greenwood Encyclopedia of American Institutions. Westport, CT: Greenwood Press, 1977.

Finn, Michael G. "Understanding the Higher Unemployment Rate of Women Scientists and Engineers." *The American Economic Review*, December 1983, 1137-40.

Fisher, Suzy. "Nora Stanton Barney, First U.S. Woman CE, Dies at 87." *Civil Engineering* 41 (April 1971):87.

Fleishman, Avrom. *Figures of Autobiography: The Language of Self-writing in Victorian and Modern England*. Berkeley: University of California Press, 1981.

Florman, Samuel. "Engineering and the Female Mind." *Harper's*, February 1978, 57-64.

−. *The Existential Pleasures of Engineering*. 1976.

Foner, Philip S. *The AFL in the Progressive Era, 1910-1915*. Vol. 5 of *History of the Labor Movement*. New York: International Publishers, 1980.

−. "Black Participation in the Centennial of 1876." *The Negro History Bulletin* 39 (1976):533-38.

−. *The Policies and Practices of the American Federation of Labor, 1900-1909*. Vol. 3 of *History of the Labor Movement in the United States*. New York: International Publishers, 1964.

−. *Post-war Struggles, 1918-1920*. In *History of the Labor Movement in the United States*. New York: International Publishers, 1988.

Fores, Michael. "Technical Change and the 'Technology' Myth." *Scandinavian Economic History Review* 30, no. 3 (1982):167-88.

Foster, Edward Halsey. "This Cowboy is an Engineer: Popular Fiction, Politics and the West." *Rendezvous: Journal of Arts and Letters* 19 (Fall 1983):1-7.

Fox, Richard Wightman, and T. J. Jackson Lears, eds. *The Culture of Consumption: Critical Essays in American History, 1880-1980*. New York: Pantheon Books, 1983.

Fox, Robert, and Anna Guagnini. "Classical Values and Useful Knowledge: The Problem of Access to Technical Careers in Modern Europe." *Daedalus* 116, no. 4 (1987):153-71.

Frederickson, George M. *The Inner Civil War. Northern Intellectuals and the Crisis of the Union*. New York: Harper & Row, 1965.

Fuchs, Margot. "Like Fathers-Like Daughters: Professionalization Strategies of Women Students and Engineers in Germany, 1890s to 1940s." *History and Technology* 14, no. 1 (1997):49-64.

Furner, Mary O. *Advocacy and Objectivity*. Lexington: University Press of Kentucky, 1975.

Garber, Marjorie. *Vested Interests: Cross Dressing and Cultural Authority*. New York: Routledge, 1992.

Gardner, Robert E. "Women: The New Engineers." In *Women in Engineering: Beyond Recruitment*, eds. Mary Ott and Nancy A. Reese. Ithaca: Cornell University Press, 1975.

Goetzmann, William H. *Army Exploration in the American West, 1803-1863*. Yale Publications in American Studies, 4. New Haven: Yale University Press, 1959.

Gombrich, Ernest. *Meditations on a Hobby Horse: And Other Essays on the Theory of Art.* 1963.

Gorn, Elliott. *The Manly Art: Bare-Knuckle Prize Fighting in America*. Ithaca, NY: Cornell University Press, 1986.

Gouda, Frances. *Poverty and Political Culture: The Rhetoric of Social Welfare Netherlands and France, 1815-1854*. Lanham, MD: Roman and Littlefield, 1995.

Green, Harvey. *Fit for America: Health, Fitness, Sport, and American Society*. Baltimore: The Johns Hopkins University Press, 1986.

Gregory, Cedric E. *A Concise History of Mining*. New York: Pergamon Press, 1980.

Gutman, Herbert G. "Two Lockouts in Pennsylvania, 1873-1874." *Pennsylvania Magazine of History and Biography* 86 (July 1959):307-26.

Haas, Violet B., and Carolyn Cummings Perrucci, eds. *Women in the Scientific and Engineering Professions*. Ann Arbor: University of Michigan Press, 1984.

Haber, Samuel. *Efficiency and Uplift: Scientific Management in the Progressive Era, 1890-1920*. Chicago: University of Chicago Press, 1964.

Hacker, Sally L. "The Culture of Engineering: Woman, Workplace and Machine." *Women Studies International Quarterly* 4, no. 3 (1981):341-53.

–. "The Mathematization of Engineering: Limits on Women and the Field." In *Machina Ex Dea: Feminist Perspectives on Technology*, edited by Joan Rothschild, 38-58. New York: Pergamon Press, 1983.

–. *Pleasure, Power, and Technology: Some Tales of Gender, Engineering, and the Cooperative Workplace*. Boston: Unwin Hyman, 1989.

Halbwachs, Maurice. *The Collective Memory*. With an introduction by Mary Douglas. Trans by Francis J. and Vida Yazdi Ditter. New York: Harper & Row, 1980 [1950].

Harris, William C. *Presidential Reconstruction in Mississippi*. Baton Rouge: Lousiana State University Press, 1967.

Harrison, Robert W. *Alluvial Empire*. Little Rock: Pioneer Press, 1961.

–. "Early State Flood-Control Legislation in the Mississippi Alluvial Valley." *Journal of Mississippi History* 23 (April 1961):104-26.

–. *Levee Districts and Levee Building in Mississippi: A Study of State and Local Efforts to Control Mississippi River Floods*. 1951.

Hartman, Mary S., and Banner Lois, eds. *Clio's Consciousness Raised: New Perspectives on the History of Women*. New York: Octagon Books, 1976.

Harvie, Christopher. "'The Sons of Martha': Technology, Transport, and Rudyard Kipling." *Victorian Studies* 20 (1977):269-82.

Havermann, Ernes, and Patricia Slater West, eds. *They Went to College*. New York: Harcourt, 1952.

Herf, Jeffrey. "The Engineer as Ideologue: Reactionary Modernists in Weimar and Nazi Germany." *Journal of Contemporary History* 19 (October 1984):631-48.

Hounshell, David. "Edison and the Pure Science Ideal in America." *Science* 207 (8 February 1980):612-7.

Hughes, Thomas P. *American Genesis: A Century of Invention and Technological Enthusiasm.* New York: Penguin, 1989.

James, Laurence P., and Sandra C. Taylor. "Strong-minded Women: Desdemona Stott Beeson and Other Hard-rock Mining Entrepreneurs." *Utah Historical Quarterly* 46, no. 2 (1978):136-50.

Jelenik, Estelle. *Women Autobiographies: Essays in Criticisms.* Bloomington: Indiana University Press, 1980.

–. "Women's Autobiography and the Male Tradition." In *Women's Autobiography: Essays in Criticism,* Ed and introduction Estelle Jelink. Bloomington: Indiana University Press, 1980.

Kaplan, Louis. *A Bibliography of American Autobiographies.* Madison: University of Wisconsin Press, 1961.

Kasson, John. *Civilizing the Machine: Technology and Republican Values in America, 1776-1900.* New York: Penguin, 1977 [1976].

Kindya, Marta Navia. *Four Decades of the Society of Women Engineers.* Edited by Cynthia Knox Lang. Society of Women Engineers, n.d.

Kolodny, Annette. *The Land Before Her: Fantasy and Experience of the American Frontiers, 1630-1860.* Chapel Hill: The University of North Carolina Press, 1984.

Kotre, John. *Outliving the Self: Generativity and the Interpretation of Lives.* Baltimore: The Johns Hopkins University Press, 1984.

Kuhn, Thomas S. "Metaphor in Science." In *Metaphor and Thought,* edited by Andrew Ortony, 409-19. Cambridge: Cambridge University Press, 1979.

Kuhns, William. *The Post-Industrial Prophets: Interpretations of Technology.* New York: Weybright and Talley, 1971.

Kwolek-Folland, Angel. *Engendering Business: Men and Women in the Corporate Office, 1870-1930.* Baltimore: The Johns Hopkins University Press, 1995.

Lakoff, George, and Mark Johnson. *Metaphors We Live By.* Chicago: University of Chicago Press, 1980.

Layton Jr., Edwin T. "Mirror-image Twins: The Communities of Science And Technology in 19th-century America." *Technology and Culture* 12 (1971):562-80.

–. *The Revolt of the Engineers: Social Responsibility and the American Engineering Profession.* Baltimore: The Johns Hopkins University Press, 1986 [1971].

–. "Science, Business and the American Engineer." In *Engineers and the Social System,* Eds Robert Perrucci and Joel E. Gerstl, 51-72. New York: John Wiley & Sons., 1969.

–. "Veblen and the Engineers." *American Quarterly* 14 (Spring 1962):64-72.

Lears, T. J. Jackson. *No Place of Grace: Antimodernism and the Transformation of American Culture, 1880-1920.* New York: Pantheon, 1981.

Lerman, Nina E. "'Preparing for the Duties and Practical Business of Life': Technological Knowledge and Social Structure in Mid-19th-Century Philadelphia." *Technology and Culture* 38 (January 1997):31-59.

–. "The Uses of Useful Knowledge: Science, Technology, and Social Boundaries in an Industrializing City." Sally Kolhstedt, and Ellen Longino. *Osiris. Special Issue Women, Gender, and the Science Question* (1997).

Levinson, Daniel, et al. *The Seasons of Man's Life.* New York: Alfred Knopf, 1978.

Lintsen, Harry. *Ingenieur van beroep: historie, praktijk, macht en opvattingen van ingenieurs in Nederland.* Den Haag: Ingenieurspers, 1985.

Lipman-Blumen, Jean. "Toward a Homosocial Theory of Sex Roles: An Explanation of the Sex Segregation of Social Institutions." *Signs* 1 (Spring 1976):15-31.

Lundgreen, Peter. "Engineering Education in Europe and the U.S.A., 1750-1930: The Rise to Dominance of School Culture and the Engineering Professions." *Annals of Science* 47 (1990):33-75.

McClintock, Anne. *Imperial Leather: Race, Gender, and Sex in Colonial Contest.* New York: Routledge, 1995.

McGaw, Judith, ed. *Early American Technology: Making and Doing Things from Colonial Times to 1850.* Chapel Hill: University of North Carolina Press, 1994.

McGaw, Judith A. "No Passive Victims, No Separate Spheres." In *In Context: The History and the History of Technology*, Eds Stephen H. Cutliffe and Robert C. Post, 172-80. Bethlehem: Lehigh University Press, 1989.

–. "Review Essay. Inventors and Other Great Women: Towards a Feminist History of Technological Luminaries." *Technology and Culture* 38 (January 1997):214-31.

–. "Review Essay: Women and the History of American Technology." *Signs* 7 (Summer 1982):798-828.

–. "Technological Change and Women's Work: Mechanization in the Berkshire Paper Industry, 1820-1855." In *Dynamos and Virgins Revisited: Women and Technological Change in History*, 77-99. Metuchen, NJ: Scarecrow Press, 1979.

–. *Most Wonderful Machine: Mechanization and Social Change in Berkshire Paper Making, 1801-1885.* Princeton: Princeton University Press, 1987.

McGovern, James R. "David Graham Philips and the Virility Impulse of the Progressives." *New England Quarterly* 39 (1966):334-55.

McIlwee, Judith S., and J. Gregg Robinson. *Women in Engineering: Gender, Power and Workplace Culture.* Albany: State University of New York Press, 1992.

Mack, Pamela. "What Difference Has Feminism Made to Engineering in the Twentieth Century?" In *Science, Medicine, Technology: The Difference Feminism Has Made*, eds. Londa Schiebinger, Elizabeth Lunbeck, and Angela N. H. Creager. Chicago: Chicago University Press, Forthcoming.

McMahon, A. Michal. *The Making of a Profession: A Century of Electrical Engineering in America.* New York: The Institute of Electrical and Electronics Engineers Press, 1984.

McMath, Robert, Jr., et al. *Engineering the New South. Georgia Tech, 1885-1985.* Athens: The University of Georgia Press, 1985.

Man, Paul de. "Autobiography as De-facement." *MLN* 94 (December 1979):919-30.

Manuel, Frank E. *The Prophets of Paris.* Cambridge, MA: Harvard University Press, 1962.

Marvin, Carolyn. *When Old Technologies Were New: Thinking About Communications in the Late Nineteenth Century.* New York: Oxford University Press, 1988.

Marx, Leo. *The Machine in the Garden.* New York: Oxford University Press, 1964.

Massa, Ann. "Black Women in the 'White City'" *Journal of American Studies* 8 (1974):319-37.

Matthews, Alva T. "Emily W. Roebling, One of the Builders of the Bridge." In *Bridge to the Future: The Celebration of the Brooklyn Bridge.* eds. Margaret Latimer, Brooke Hindle, and Melvin Kranzberg. 63-70. New York: New York Academy of Sciences, 1984.

Mayr, Otto. "The Science-Technology Relationship as a Historiographic Problem." *Technology and Culture*, October 1976, 663-73.

Meiksins, Peter F. "Beyond the Boundary Question." *New Left Review* (1986):101-20.

–. "Engineers in the United States: A House Divided." In *Engineering Labour: Technical Workers in a Comparative Perspective*, eds. Peter Meiksins and Smith Chris, 61-97. London: Verso, 1996.

–. "Productive and Unproductive Labor and Marxs Theory of Class." *Review of Radical Political Economics* 13, no. 3 (1981):32-42.

–. "Professionalism and Conflict: The Case of the American Association of Engineers." *Journal of Social History* 19 (Spring 1986):403-22.

–. "'The Revolt of the Engineers' Reconsidered." *Technology and Culture* 29, no. 2 (1988):219-46.

–. *Science in the Labor Process: Engineers as Workers.* In *Professionals as Workers: Mental Labor in Advanced Capitalism*, edited by Charles Derber, 121-40. Boston: G. K. Hall, 1982.

–. "Scientific Management and Class Relations – a Dissenting View." *Theory and Society* 13 (March 1984):177-209.

Meiksins, Peter F., and James M. Watson. "Professional Autonomy and Organizational Constraint – the Case of Engineers." *Sociological Quarterly* 30, no. 4 (1989):561-85.

–. "What Do Engineers Want? Work Values, Job Rewards, and Job Satisfaction." *Science, Technology, & Human Values* 16 (Spring 1991):140-72.

Meiksins, Peter F., and Chris Smith. eds. *Engineering Labour: Technical Workers in Comparative Perspective.* London: Verso, 1996.

Melosh, Barbara. *The Physicians' Hand: Work Culture and Conflict in American Nursing.* Philadelphia: Temple University Press, 1982.

Merritt, Raymond H. *Engineering in American Society, 1850-1875.* Lexington: The University Press of Kentucky, 1969.

Merton, Robert K. "Machine, the Worker, and the Engineer." *Science* 105 (24 January 1947):79-84.

Mikoletzky, Juliane. "Unintended Consequence: Women's Entry Into Engineering Education." *History and Technology* 14 (January 1997):31-48.

Miller, Donald L. *City of the Century. The Epic of Chicago and the Making of America.* New York: Touchstone, 1996.

Misa, Thomas J. *A Nation of Steel: The Making of Modern America, 1865-1925.* Baltimore: The Johns Hopkins University Press, 1995.

Molella, Arthur P. "The First Generation: Usher, Mumford, and Giedion." In *In Context: The History and the History of Technology*, Stephen H. Cutliffe, and Robert C. Post, 88-105. Bethlehem: Lehigh University Press, 1989.

–. "The Museum That Might Have Been: The Smithsonian's National Museum of Engineering and Industry." *Technology and Culture* 32 (April 1991):237-63.

Molella, Arthur P., and Nathan Reingold. "Theorists and Ingenious Mechanics: Joseph Henry Defines Science." *Science Studies* 3 (1973):322-35.

Montgomery, David. *The Fall of the House of Labor: The Workplace, the State, and American Labor Activism, 1865-1925.* Cambridge: Cambridge University Press, 1987.

–. "Workers' Control of Machine Production in the Nineteenth Century." In *Workers' Control in America: Studies in the History of Work, Technology, and Labor Struggles*, 9-31. Cambridge: Cambridge University Press, 1979.

Morrell, Jack, and Arnold Thackray. *Gentlemen of Science. Early Years of the British Association for the Advancement of Science.* Oxford: Clarendon Press, 1981.

Mukerji, Chandra and Michael Schudson, eds. *Rethinking Popular Culture: Contemporary Thought in Cultural Studies.* Berkeley: University of California Press, 1991.

Multhauf, Robert P. "Some Observations on the Historiography of the Industrial Revolution." In *In Context: History and the History of Technology.*, Stephen H. Cutcliffe, and Robert C. Post. Bethlehem: Lehigh University Press, 1989.

Mumford, Lewis. "If Engineers Were Kings." *The Freeman* 4 (23 November 1921):261-62.

–. *Technics and Civilization.* New York: Harcourt, Brace, 1934.

–. "Technics and the Nature of Man." *Technology and Culture* 7 (1966).

National Association of State Universities and Land-Grant Colleges. *Leadership and Learning: An Interpretative History of Historically Black Land-Grant Colleges and Universities.* 1993.

Naumann, Francis. "A Lost American Futurist." *Art in America*, April 1994.

Nesbit, Molly. "Ready-Made Originals: The Duchamp Model." *October* 37 (Summer 1986):53-64.

Newman, Louise Michele, ed. *Men's Ideas/Women's Realities: Popular Science, 1870-1915.* With a foreword by Ruth Hubbard. New York: Pergamon Press, 1985.

Nieman, Donald G. *African Americans and Education in the South, 1865-1900.* New York: Garland, 1994.

Noble, David F. *America by Design: Science, Technology and the Rise of Corporate Capitalism.* New York: Cambridge University Press, 1977.

–. *Forces of Production: A Social History of Industrial Automation.* New York: Oxford University Press, 1984.

–. *A World Without Women: The Christian Clerical Culture of Western Science.* New York: Alfred A. Knopf, 1992.

Nora, Philip. "Between Memory and History." *Representations* 26 (1989):7-22.

O'Brien, Sharon. "Becoming Noncanonical." *American Quarterly* 40 (March 1988).

O'Connell Jr., Charles F. "The Corps of Engineers and the Rise of Modern Management, 1827-1856." In *Military Enterprise and Technological Change: Perspectives on the American Experience*, edited by Merritt Roe Smith, 87-116. Cambridge, MA: MIT Press, 1987 [1985].

Oakley, Ann. *Sex, Gender and Society.* London: Temple Smith, 1972.

Oehlschlaeger, Fritz H. "Civilization as Emasculation: The Threatening Role of Women in the Frontier Fiction of Harold Bell Wright and Zane Grey." *The Midwest Quarterly* 22 (Summer 1981):346-60.

Oldenziel, Ruth. "Artistic Crossings: The Ford (plant) at River Rouge, 1927-1939." In *High Brow Meets Low Brow: American Culture as an Intellectual Concern*, edited by Rob Kroes, 37-60. Netherlands American Studies Association Series. Amsterdam: Free University Press, 1988.

Oldenziel, Ruth, Karin Zachmann, and Annie Canel. *Crossing Boundaries: The History of Women Engineers in the Comparison.* London: Harwood Academic Press, Forthcoming.

Orjollet, Jean-François. "Individu, Type, Règle: Kipling et Ses Engineers." *CVE* 18 (Novembre 1983):59-67.

Ortony, Andrew, ed. *Metaphor and Thought.* Cambridge: Cambridge University Press, 1979.

Ott, Mary and Nancy A. Reese (eds.). *Women in Engineering: Beyond Recruitment.* Ithaca: Cornell University Press, 1975.

Parry, Ann. "Imperialism in 'The Bridge-Builders': Metaphor or Reality?" *Kipling Journal* 60 (March 1986):12-22.

–. "Imperialism in 'The Bridge-Builder' The Builders of the Bridge and the Future of the Raj." *Kipling Journal* 60 (June 1986):9-16.

Perrucci, Carolyn C. "Minority Status and the Pursuit of Professional Careers: Women in Science and Engineering." *Social Forces* 49 (December 1970):245-59.

–. "Engineering and the Class Structure." In *The Engineers and the Social System,* Eds Robert Perrucci and Joel Gerstl. New York: John Wiley and Sons, 1969.

Pfaffenberger, Bryan. "Fetishised Objects and Humanised Nature: Toward an Anthropology of Technology." *Man* 23 (June 1988):236-52.

Pickthall, Marjorie Lowry Christie. *The Bridge: A Story of the Great Lakes.* New York: Century, 1922.

Pleck, Elizabeth H., and Joseph H. Pleck, eds. *The American Man.* Englewood Cliffs, NJ: Prentice Hall, 1980.

Pleck, Joseph. *The Myth of Masculinity.* Cambridge, MA: MIT Press, 1981.

Post, Robert C. "Reflections of American Science and Technology at the New York Crystal Palace Exhibition of 1853." *Journal of American Studies* 17 (1983):337-56.

Post, Robert C. ed. *1876: A Centennial Exhibition. A Catalogue to the Exhibit.* Washington, D.C.: Smithsonian Institution, 1976.

Pugh, David G. *Sons of Liberty: The Masculine Mind in 19th Century America.* Westport, CT: Greenwood Press, 1983.

Pursell, Carroll. "'Am I a Lady or an Engineer?' The Origins of the Women's Engineering Society in Britain, 1918-1940." *Technology and Culture* 34 (January 1993):78-97.

–. "The Construction of Masculinity and Technology." *Polhem,* no. 11 (1993):206-19.

–. "The Cover Design: Women Inventors in America." *Technology and Culture* 22 (1981):545-48.

–. *The Machine in America: A Social History of Technology.* Baltimore: The Johns Hopkins University Press, 1995.

–. "Seeing the Invisible. New Perceptions in the History of Technology." *Icon* 1 (1995):9-15.

–. "Toys, Technology and Sex Roles in America, 1920-1940." In *Dynamos and Virgins Revisited: Women and Technological Change,* ed. Martha Moore Trescott, 252-67. Methuen, NJ: Scarecrow Press, 1979.

Quimby, Ian. *Apprenticeship in Colonial Philadelphia.* New York, 1985 [1963].

Radway, Janice A. *Reading the Romance: Women, Patriarchy, and Popular Literature.* Chapel Hill: The University of North Carolina Press, 1984.

Reskin, Barbara F., and Polly A. Phipps. "Women in Male-Dominated Professional and Managerial Occupations." In *Women Working: Theories and Facts in Perspective,* eds. Ann H. Stromberg, and Shirley Harkness. Mountain View, CA: Mayfield, 1988.

Reskin, Barbara F., and Patricia A. Roos. *Job Queues, Gender Queues: Explaining Women's Inroads Into Male Occupations.* Philadelphia: Temple University Press, 1990.

Reynolds, Terry S. "Defining Professional Boundaries: Chemical Engineering in the Early Twentieth Century." *Technology and Culture* 27 (October 1988):694-716.

–. ed. *The Engineer in America: A Historical Anthology from Technology and Culture.* Chicago: Chicago University Press, 1991.

Roark, James L. *Masters Without Slaves: Southern Planters in the Civil War and Reconstruction.* New York: Norton, 1977.

Robin, Stanley S. "The Female in Engineering." In *The Engineers and the Social System,* Eds Robert C. Perrucci and Joel Gerstl, 203-18. New York: John Wiley and Sons, 1969.

Robinson, J. Gregg, and Judith S. McIlwee. "Women in Engineering: A Promise Unfulfilled?" *Social Problems* 36 (1989):455-72.

Rodgers, Daniel T. *Contested Truths. Keywords in American Politics Since Independence.* New York: Basic Books, 1987.

–. *The Work Ethic in Industrial America, 1850-1920.* Chicago: The University of Chicago Press, 1974.

Roe Smith, Merritt, ed. *Military Enterprise and Technological Change: Perspectives on the American Experience.* Cambridge, MA: MIT Press, 1987 1985

Rollins, Richard M. "Words as Social Control: Noah Webster and the Creation of the 'American Dictionary'" *American Quarterly* 28 (1976):415-30.

Rolt, L. T. C. *A Short History of Machine Tools.* Cambridge, MA: MIT Press, 1965.

Romein, Jan. *De biografie: een inleiding.* Amsterdam: Ploegsma, 1946.

Rorabaugh, W. J. *The Craft Apprentice: From Franklin to the Machine Age in America.* Oxford: Oxford University Press, 1986.

Rose, Mark H. "Science as an Idiom in the Domain of Technology." *Science and Technology Studies* 5, no. 1 (1987):3-11.

Rosenberg, Bernard. "A Clarification of Some Veblenian Concepts." *American Journal of Economics and Sociology* 12 (January 1953):179-87.

Rosenberg, Nathan. *Inside the Black Box: Technology and Economics.* New York: Cambridge University Press, 1982.

–. *Perspectives on Technology.* New York: Cambridge University Press, 1976.

Rosenberg, Rosalyn. *Beyond Separate Spheres: Intellectual Roots of Modern Feminism.* New Haven: Yale University, 1983.

Ross, Dorothy. "Development of the Social Sciences." Eds Alexandra Oleson and John Voss. In *The Organization of Knowledge in Modern America, 1860-1920.* Baltimore: The Johns Hopkins University Press, 1979.

Rossiter, Margaret W. *Women Scientists in America. Before Affirmative Action, 1940-1972.* Baltimore: The Johns Hopkins University Press, 1995.

–. *Women Scientists in America. Struggles and Strategies to 1940.* Baltimore: The Johns Hopkins University Press, 1982.

Rothschild, Joan (ed.). *Machina Ex Dea: Feminist Perspectives on Technology.* Athene Series. New York: Pergamon Press, 1983.

Rotundo, E. Anthony. *American Manhood: Transformations in Masculinity from the Revolution to the Modern Era.* New York: Basic Books, 1993.

–. "Body and Soul: Changing Ideals of American Middle-Class Manhood, 1770-1920." *Journal of Social History* (1983):22-38.

–. "Boy Culture: Middle-class Boyhood in Nineteenth-century America." In *Meanings for Manhood: Constructions of Masculinity in Victorian America*, 15-36. Chicago: The University of Chicago Press, 1990.

Rudwick, Elliott M., and August Meier. "Black Man in the 'White City': Negroes and the Columbian Exposition, 1893." *Phylon* 26 (1965):354-61.

Rydell, Robert W. *All the World's a Fair: Visions of Empire at American International Expositions, 1876-1916*. Chicago: The University of Chicago Press, 1984.

–. "The Fan Dance of Science; American World's Fairs in the Great Depression." *Isis* 76, no. 284 (1985):525-42.

–. "The Literature of International Expositions." In *The Books of the Fairs. Material About the World's Fairs, 1834-1916, in the Smithsonian Institution Libraries*, 1-62. Chicago: American Library Association, 1992.

–. *World of Fairs: The Century-of-Progress Expositions*. Chicago: University of Chicago Press, 1993.

Salembier, Olive. "Women Engineers." In *Women in Engineering: Bridging the Gap Between Society and Technology*, George Bugliarello, et al. (eds.). Chicago: University of Illinois, 1971.

Schiebinger, Londa, Elizabeth Lunbeck, and Angela N. H. Creager eds. *Science, Medicine, and Technology*. Chicago: Chicago University Press, Forthcoming.

Scott, Joan W. "Gender: A Useful Category of Historical Analysis." *Journal of American History* 91 (1986):1053-75.

Scranton, Philip. "Learning Manufacture: Education and Shop-floor Schooling in the Family Firm." *Technology and Culture* 27 (January 1986):40-62.

Seeley, Bruce E. "Research, Engineering, and Science in American Engineering Colleges, 1900-1960." *Technology and Culture* 34 (1993):344-86.

–. "SHOT, the History of Technology and Engineering Education." *Technology and Culture* 36 (October 1995):739-72.

Segal, Howard P. *Technological Utopianism in American Culture*. Chicago: The University of Chicago Press, 1985.

Sicherman, Barbara. "Sense and Sensibility: A Case Study of Women's Reading in Late-Victorian America." In *Reading in America: Literature and Social History*, ed. Cathy N. Davidson, 201-25. Baltimore: The Johns Hopkins University Press, 1989.

Sinclair, Bruce. *A Centennial of the American Society of Mechanical Engineers, 1880-1980*. Toronto: University of Toronto Press, 1980.

–. "Inventing a Genteel Tradition: MIT Crosses the River." In *New Perspectives on Technology and American Culture*, edited by Bruce Sinclair, 1-18. American Philosophical Society Library Publication series no. 12. Philadelphia: American Philosophical Society, 1986.

–. "Local History and National Culture: Notions on Engineering Professionalism in America." *Technology and Culture* 27 (October 1986):683-93.

–. "Notions on Engineering Professionalism in America." *Technology and Culture* 27 (October 1986):683-93.

–. *Philadelphia's Philosopher Mechanics: A History of the Franklin Institute, 1824-1865.* Baltimore: The Johns Hopkins University Press, 1974.

Smith, Frank E. *Yazoo River.* Jackson: University Press of Mississippi, 1954.

Smith, Sidonie. *A Poetics of Women's Autobiography. Marginality and the Fictions of Self-representation.* Bloomington: Indiana University Press, 1987.

Smith-Rosenberg, Carroll. "The Female World of Love and Ritual: Relations Between Women in Ninetieth-century America." *Signs* 1 (Autumn 1975):1-29.

Smithsonian Year. Annual Report of the Smithsonian. Washington, D.C.: Smithsonian Institution Press, 1981.

Spees, Pam. "Retired Mechanical Engineer Beatina Alexander, 81, Fights for Cleaner Environment." *Flair,* 14 December 1989, A6-7.

Spence, Clark C. *Mining Engineers & the American West: The Lace-Boot Brigade, 1849-1919.* New Haven: Yale University Press, 1970.

Spivey, Donald. *Schooling for the New Slavery: Black Industrial Education, 1868-1915.* Westport, CT: Greenwood Press, 1978.

Stage, Sarah, and Virginia B. Vincenti, eds. *Rethinking Home Economics. Women and the History of a Profession.* Ithaca: Cornell University Press, 1997.

Stanley, Autumn. *Mothers and Daughters of Invention: Notes for a Revised History of Technology.* Methuen, NJ: Scarecrow Press, 1993.

–. "The Patent Office as Conjurer: The Vanishing Lady Trick in a Nineteenth-century Historical Source." Eds Barbara Wright Drygulski et al. In *Women, Work, and Technology,* 118-36. Michigan: The University of Michigan Press, 1987.

Staudenmaier, John M. "Technology's Storytellers: Reweaving the Human Fabric." Cambridge, MA: MIT Press, 1985.

Stearns, Peter N. *Bed Man: Males in Modern Society.* New York: Holms and Maier, 1979.

Stepan, Nancy Leys. "Race and Gender: The Role of Analogy in Science." *Isis* 77 (1986):261-77.

Stilgoe, John R. "Moulding the Industrial Zone Aesthetic: 1880-1929." *Journal of American Studies* 16 (April 1982).

Stine, Jeffrey K. "Nelson P. Lewis and the City Efficient: The Municipal Engineer in City Planning During the Progressive Era." Essays in Public Works History, no. 11. Chicago: Public Works Historical Society, 1981.

Susman, Warren I. *Culture as History: The Transformation of American Society in the Twentieth Century.* New York: Pantheon Books, 1984.

–. "The People's Fair: Cultural Contradictions of a Consumer Society." In *Culture as History: The Transformation of American Society in the Twentieth Century,* 211-29. New York: Pantheon, 1984.

Taylor, William R., and Christopher Lasch. "Two 'Kindred Spirits': Sorority and Family in New England, 1839-1846." *New England Quarterly* 36 (1963):25-36.

Thelen, David. "Memory and American History." *Journal of American History* 75 (March 1989):1117-29.

Tichi, Cecelia. *Shifting Gears: Technology, Literature, Culture in Modernist America.* Chapel Hill: The University of North Carolina Press, 1987.

Tolson, Andrew. *The Limits of Masculinity.* New York: Harper and Row, 1977.

Trachtenberg, Alan. *The Brooklyn Bridge: Fact and Symbol.* Chicago: University of Chicago Press, 1979 [1965].

Travers, Tim. *Samuel Smiles and the Victorian Work Ethic.* Garland Series of Outstanding Dissertations. New York: Garland Publishing, 1987 [1970].

Trescott, Martha Moore, ed. *Dynamos and Virgins Revisited: Women and Technological Change in History.* Metuchen, NJ: Scarecrow Press, 1979.

–. "Lillian Moller Gilbreth and the Founding of Modern Industrial Engineering." In *Machina Ex Dea: Feminist Perspectives on Technology,* edited by Joan Rothschild, 23-37. New York: Pergamon Press, 1983.

–. "Women Engineers in History: Profiles in Holism and Persistence." In *Women in Scientific and Engineering Professions,* Eds Violet B. Haas and Carolyn C. Perrucci, 181-205. Ann Arbor: The University of Michigan Press, 1984.

Tsujimoto, Karen. *Images of America: Precisionist Painting and Modern Photography.* Seattle and London: University of Washington Press, 1982.

Turner, Edna M. "Distinguished Alunmae." *The Pratt Alumnus,* March 1951.

Tyler, Robert L. "The I.W.W. and the Brainworkers." *American Quarterly* 15 (Spring 1963).

Vare, Ethlie Ann, and Greg Ptacek. *Mothers of Invention: From the Bra to the Bomb, Forgotten Women and Their Unforgettable Ideas.* New York: William Morrow, 1987.

Vetter, Betty. "Women Scientists and Engineers: Trends in Participation." *Science* 214 (18 December 1981):1313-21.

Vetter, Betty, and Eleanor L. Babco. *Professional Women and Minorities: A Manpower Data Resource Service.* Washington, D.C.: Commission on Professionals in Science and Technology, 1989 (6th edition).

Vitale, John C. "The Great Quebec Bridge Disaster." *Consulting Engineer* 38 (February 1967):92-5.

Wajcman, Judy. "Sally Hacker, Pleasure, Power, and Technology [and Hacker] 'Doing It the Hard Way'" Review Essay. *Signs* 17 (winter 1992):478-80.

Warner, Deborah J. "The Woman's Pavilion." In *1876: A Centennial Exhibition,* Robert C. Post. Washington, D.C.: Smithsonian Institution, 1976.

–. *Women in Science in Nineteenth-century America.* Washington, D.C.: Smithsonian Institution, 1978.

–. "Women Inventors at the Centennial." In *Dynamos and Virgins Revisited,* edited by , 102-19. Metuchen, NJ: Scarecrow Press, 1979.

Way, Peter. *Common Labor: Workers and the Digging of North American Canals, 1780-1860.* Baltimore: The Johns Hopkins University Press, 1993.

Weigold, Marilyn. *The Silent Builder: Emily Warren Roebling and the Brooklyn Bridge.* Port Washington, NY: Associated Faculty Press, 1984.

Weimann, Jeanne Madeline. *The Fair Women: The Story of the Woman's Building, World's Columbian Exposition Chicago 1893.* Chicago: Academy Chicago, 1981.

–. "The Great 1893 Woman's Building: Can We Measure Up in 1992." *MS Magazine* 41 (March 1983):65-7.

Wiebe, Robert H. *The Search for Order, 1877-1920.* New York: Hill and Wang, 1967.

Wilentz, Sean. *Chants Democratic. New York City & the Rise of the American Working Class, 1788-1850.* New York: Oxford University Press, 1984.

Williams, Raymond. *Keywords. A Vocabulary of Culture and Society*. New York: Oxford University Press, 1976.

–. *Keywords: A Vocabulary of Culture and Society*. New York: Oxford University Press, 1983 [1976].

Willis, Paul. "Masculinity, Wage Form, and Factory Labor." In *Working Class Culture*, Eds Clarke John et al, 185-98. London: Hutchinson, 1979.

Wilson, Christopher P. "The Rhetoric of Consumption: Mass-market Magazines and the Demise of the Gentle Reader, 1880-1920." In *Culture of Consumption: Critical Essays in American History, 1880-1980*, Eds Richard Wightman Fox and T. J. Jackson Lears, 39-64. New York: Pantheon Books, 1983.

Wilson, Richard, Dianne H. Pilgrim, and Tashjian. *The Machine Age in America, 1918-1941*. Catalogue Brooklyn Museum. New York: Brooklyn Museum, 1986.

Winner, Langdon. *Autonomous Technology: Technics-out-of Control as a Theme of Political Thought*. Cambridge, MA: MIT Press, 1989 [1977]

Wise, George. "Heat Transfer Research in General Electric, 1910-1960." History of Heat transfer in honor of the 50th anniversary of the ASME heat transfer division. New York: ASME, 1988.

Wisely, W. H. *The American Civil Engineer, 1852-1974: The History, Traditions and the Development of the American Society of Civil Engineers*. New York: American Society of Civil Engineers, 1974.

Wright, Barbara Drygulski, ed. *Women, Work and Technology: Transformations*. Ann Arbor: University of Michigan Press, 1987.

Yacovone, Donald. "Abolitionists and the 'Language of Fraternal Love'" In *Meanings for Manhood: Constructions of Masculinity in Victorian America*, Mark C. and Clyde Griffen Carnes, 85-95. Chicago: University of Chicago Press, 1990.

Yates, JoAnne. *Control Through Communication: The Rise of System in American Management*. Baltimore: The Johns Hopkins University, 1989.

Zabel, Barbara. "The Machine as Metaphor, Model, and Microcosm: Technology in American Art, 1915-1930." *Arts Magazine* 57 (12 December 1982):100-05.

Zachmann, Karin. "'Involvement with Technology Does not Harm Their Charm and Femininity'. Mobilization, Modernization, and Ambivalence of Women Engineers in East Germany." In *Crossing Boundaries: History of Women Engineers in Cross-Cultural Comparison*, eds. Ruth Oldenziel, Karin Zachmann, and Annie Canel. London: Harwood Academic Press, Forthcoming.

Ziolkowski, Theodore. "The Existential Anxieties of Engineering." *The American Scholar* 53 (Spring 1984):197-218.

Zunz, Olivier. *Making America Corporate, 1870-1920*. Chicago: The University of Chicago Press, 1990.

Unpublished
Theses and Papers

Bever, Marilynn A. *The Women at MIT, 1871-1941: Who They Were, What They Achieved*. 2 Vols. BA Thesis MIT. 1976.

Bix, Amy Sue. *Inventing Ourselves Out of Jobs: America's Depression Era Debate Over Technological Unemployment*. Ph.D. Diss., Johns Hopkins University. 1994.

Bristol-Kagan, Leigh. *Chinese Migration to California, 1851-1882: Selected Industries of Work, the Chinese Institutions and the Legislative Exclusion of a Temporary Labor Force.* Ph.D. Diss., Harvard University. 1982.

Cooper, Gail. *The Manufactured Weather: The History of Airconditioning in the United States, 1906-1979.* Ph.D. Diss.

Crane, Diana. "In Search of Technology: Definitions of Technology in Different Fields – an Essay Review." Paper Presented at Mellon Foundation Seminar on Technology and Culture . 1987.

Darney, Virginia Grant. *Women and World's Fairs: American International Expositions, 1876-1904.* Ph.D. Diss., Emory University. 1982.

Goldstein, Carolyn. *Mediating Consumption: Home Economics and American Consumers, 1900-1940.* Ph.D. Diss., University of Delaware. 1994.

Jordan, John M. *Technic and Ideology: The Engineering Ideal and American Political Culture.* Dissertation, the University of Michigan. 1989.

Lerman, Nina E. *From 'Useful Knowledge' to 'Habits of Industry': Gender, Race and Class in Nineteenth Century Technical Education.* Ph.D. Diss, University of Pennsylvania. 1993.

Lubar, Steve. "Don't Fold, Spindle, or Mutilate: A Cultural History of the Punchcard." Paper Presented at the Smithsonian Institution. Washington, D.C. 1990.

Meiksins, Peter. "Engineers and Managers: An Historical Perspective on an Uneasy Relationship." Paper Presented to the American Sociological Association Meetings, San Francisco, CA. 1989.

Oldenziel, Ruth. "Contesting the Good Part: Kipling's Sons of Martha and Definitions of Labor." Paper Presented at Department of Comparative Literature Seminar, Yale University. 11 April 1990.

–. *Gender and the Meanings of Technology: Engineering in the U.S., 1880-1945.* Ph.D. Diss., Yale University. 1992.

Rhees, David J. *The Chemists' Crusade: The Rise of an Industrial Science in Modern America, 1907-1922.* Ph.D. Diss, University of Philadelphia. 1987.

Seeley, Bruce E. "Changing Patterns of Research in American Engineering Colleges: The Social Dimension of the Rise of Engineering Science." Paper Presented at Scientific and Technological Development in the Nineteenth and Twentieth Century. Eindhoven, 1990.

Travers, Tim. *Samuel Smiles and the Victorian Work Ethic.* Garland Series of Outstanding Dissertations. New York: Garland Publishing, 1987 [1970].

Turner, Edna M. *Education of Women for Engineering in the United States, 1885-1952.* Ph.D. Diss., New York University, 1954.

Turner, Matt L. "Contesting the Good Part: Kipling's Sons of Martha." Paper Presented at Department of Comparative Literature Seminar, Yale University. New Haven, CT, 11 April 1990.

Index

African Americans, 23-24, 40, 69-70, 73, 102
  as engineers, 63, 69-70
academic engineers, 62. *See also* formal education
aeronautical engineering, 158
Ainsworth, Danforth H., 66, 97, 98, 100, 108
Alcott, Louisa May, 120
Alexander, Archie, 91
Alexander's Bridge (Willa Cather), 139-40
Allison, Emma, 32, 45
American Association of Engineers (AAE), 77, 79, 81, 85, 115
American Federation of Technical Engineers, 87
American Institute of Chemical Engineers, 73, 170
American Institute of Electrical Engineers (AIEE), 72, 168
American Institute of Mining and Metallurgical Engineers, 167-68
American Institute of Mining Engineers (AIME), 112
American Society of Civil Engineers (ASCE), 71, 94
  exclusionary policies of, 71, 72, 81, 148, 168-69
  women and, 148, 168-69
American Society of Mechanical Engineers (ASME), 108, 109, 111, 116
Anthony, Susan B., 32, 40, 152
anthropologists, 26-27
  and inventions, 26-27, 38-40, 49

Antonova, Helen A., 168
"applied science" (term), 63, 64, 70
artists, 120. *See also* modern art
associations. *See* engineering societies
autobiographies (of male engineers), 16, 91-106, 119
  disembodied language of, 91, 93, 96-100
  and middle-class identity, 16, 92, 93
  workers' place in, 92, 100-102, 108-9, 113

Babcock, Garrison, 62
Bache, Alexander Dallas, 22
Baker, Elizabeth, 47
Baltimore & Ohio Railroad, 55, 165, 167
Barksdale, Hamilton, 59-60
Barney, Nora Stanton Blatch. *See* Blatch, Nora Stanton
Bates, Onward, 103-4
Beard, Charles A., 42, 43
Beard, Mary Ritter, 17, 28, 181, 182-84, 190
Bell, Alexander Graham, 72
Bensel, John A., 72, 74
Bigelow, Jacob, 23, 42, 64
Bix, Amy, 46, 48
Blanchard, Helen, 28
Blatch, Harriet Stanton, 148
Blatch, Nora Stanton, 148, 149, 153-54, 157-58, 161, 168-69, 173
Boas, Franz, 49
"borrowed identity," 150-51. *See also* family ties